THE RAILROAD BUILDERS

EXTRA-ILLUSTRATED EDITION

∴

VOLUME 38
THE CHRONICLES
OF AMERICA SERIES
ALLEN JOHNSON
EDITOR

GERHARD R. LOMER
CHARLES W. JEFFERYS
ASSISTANT EDITORS

JAMES J. HILL

Photograph by Pach Bros., New York.

THE
RAILROAD BUILDERS

A CHRONICLE OF THE
WELDING OF THE STATES
BY JOHN MOODY

NEW HAVEN: YALE UNIVERSITY PRESS
TORONTO: GLASGOW, BROOK & CO.
LONDON: HUMPHREY MILFORD
OXFORD UNIVERSITY PRESS

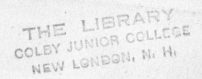

CONTENTS

ILLUSTRATIONS

THE RAILROAD BUILDERS

▽

CHAPTER I

A CENTURY OF RAILROAD BUILDING

THE United States as we know it today is largely the result of mechanical inventions, and in particular of agricultural machinery and the railroad. One transformed millions of acres of uncultivated land into fertile farms, while the other furnished the transportation which carried the crops to distant markets. Before these inventions appeared, it is true, Americans had crossed the Alleghanies, reached the Mississippi Valley, and had even penetrated to the Pacific coast; thus in a thousand years or so the United States might conceivably have become a far-reaching, straggling, loosely jointed Roman Empire, depending entirely upon its oceans, internal watercourses, and imperial highways for such economic and political integrity as it might

1

achieve. But the great miracle of the nineteenth century — the building of a new nation, reaching more than three thousand miles from sea to sea, giving sustenance to more than one hundred million free people, and diffusing among them the necessities and comforts of civilization to a greater extent than the world had ever known before — is explained by the development of harvesting machinery and of the railroad.

The railroad is sprung from the application of two fundamental ideas — one the use of a mechanical means of developing speed, the other the use of a smooth running surface to diminish friction. Though these two principles are today combined, they were originally absolutely distinct. In fact there were railroads long before there were steam engines or locomotives. If we seek the real predecessor of the modern railroad track, we must go back three hundred years to the wooden rails on which were drawn the little cars used in English collieries to carry the coal from the mines to tidewater. The natural history of this invention is clear enough. The driving of large coal wagons along the public highway made deep ruts in the road, and some ingenious person began repairing the damage by laying wooden planks in the furrows.

The coal wagons drove over this crude roadbed so successfully that certain proprietors started constructing special planked roadways from the mines to the river mouth. Logs, forming what we now call "ties," were placed crosswise at intervals of three or four feet, and upon these supports thin "rails," likewise of wood, were laid lengthwise. So effectually did this arrangement reduce friction that a single horse could now draw a great wagon filled with coal — an operation which two or three teams, lunging over muddy roads, formerly had great difficulty in performing. In order to lengthen the life of the road, a thin sheeting of iron was presently laid upon the wooden rail. The next improvement was an attempt to increase the durability of the wagons by making the wheels of iron. It was not, however, until 1767, when the first rails were cast entirely of iron with a flange at one side to keep the wheel steadily in place, that the modern roadbed in all its fundamental principles made its appearance. This, be it observed, was only two years after Watt had patented his first steam engine, and it was nearly fifty years before Stephenson built his first locomotive. The railroad originally was as completely dissociated from steam propulsion as was the ship. Just as vessels had existed for ages before the introduction of mechanical power, so the railroad had been a familiar sight in the mining districts of England for at least two centuries before the invention of Watt really gave it wings and turned it to wider uses. In this respect the progress of the railroad resembles that of the automobile, which had existed in crude form long before the invention of the gasoline engine made it practically useful.

In the United States three new methods of transportation made their appearance at almost the same time — the steamboat, the canal boat, and the rail car. Of all three, the last was the slowest in attaining popularity. As early as 1812 John Stevens, of Hoboken, aroused much interest and more amused hostility by advocating the building of a railroad, instead of a canal, across New York State from the Hudson River to Lake Erie, and for several years this indefatigable spirit journeyed from town to town and from State to State, in a fruitless effort to push his favorite scheme. The great success of the Erie Canal was finally hailed as a conclusive argument against all the ridiculous claims made in favor of the railroad and precipitated a canal mania which spread all over the country.

Yet the enthusiasts for railroads could not be discouraged, and presently the whole population divided into two camps, the friends of the canal, and the friends of the iron highway. Newspapers acrimoniously championed either side; the question was a favorite topic with debating societies; public meetings and conventions were held to uphold one method of transportation and to decry the other. The canal, it was urged, was not an experiment; it had been tested and not found wanting; already the great achievement of De Witt Clinton in completing the Erie Canal had made New York City the metropolis of the western world. The railroad, it was asserted, was just as emphatically an experiment; no one could tell whether it could ever succeed; why, therefore, pour money and effort into this new form of transportation when the other was a demonstrated success?

It was a simple matter to find fault with the railroad; it has always been its fate to arouse the opposition of the farmers. This hostility appeared early and was based largely upon grounds that have a familiar sound even today. The railroad, they said, was a natural monopoly; no private citizen could hope ever to own one; it was thus a kind of monster which, if encouraged, would override all popular rights. From this economic criticism the enemies of the railroad passed to details of construction: the rails would be washed out by rains; they could be destroyed by mischievous people; they would snap under the cold of winter or be buried under the snow for a considerable period, thus stopping all communication. The champions of artificial waterways would point in contrast to the beautiful packet boats on the Erie Canal, with their fine sleeping rooms, their restaurants, their spacious decks on which the fine ladies and gentlemen congregated every warm summer day, and would insist that such kind of travel was far more comfortable than it could ever be on railroads. To all these pleas the advocates of the railroad had one unassailable argument — its infinitely greater speed. After all, it took a towboat three or four days to go from Albany to Buffalo, and the time was not far distant, they argued, when a railroad would make the same trip in less than a day. Indeed, our forefathers made one curious mistake: they predicted a speed for the railroad — a hundred miles an hour — which it has never attained consistently with safety.

If the American of today could transport himself to one of the first railroad lines built in the United

States it is not unlikely that he would side with the canal enthusiast in his argument. The rough pictures which accompany most accounts of early railroad days, showing a train of omnibus-like carriages pulled by a locomotive with upright boiler, really represent a somewhat advanced stage of development. Though Stephenson had demonstrated the practicability of the locomotive in 1814 and although the American, John Stevens, had constructed one in 1826 which had demonstrated its ability to take a curve, local prejudice against this innovation continued strong. The farmers asserted that the sparks set fire to their hayricks and barns and that the noise frightened their hens so that they would not lay and their cows so that they could not give milk. On the earliest railroads, therefore, almost any other method of propulsion was preferred. Horses and dogs were used, winches turned by men were occasionally installed, and in some cases cars were even fitted with sails. Of all these methods, the horse was the most popular: he sent out no sparks, he carried his own fuel, he made little noise, and he would not explode. His only failing was that he would leave the track; and to remedy this defect the early railroad builders hit upon a happy device. Sometimes they would fix a treadmill inside the car; two horses would patiently propel the caravan, the seats for passengers being arranged on either side. So unformed was the prevalent conception of the ultimate function of the railroad, and so pronounced was the fear of monopoly that, on certain lines, the roadbed was laid as a state enterprise and the users furnished their own cars, just as the individual owners of towboats did on the canals. The drivers, however, were an exceedingly rough lot; no schedules were observed and as the first lines had only single tracks and infrequent turnouts, when the opposing sides would meet each other coming and going, precedence was usually awarded to the side which had the stronger arm. The roadbed showed little improvement over the mine tramways of the eighteenth century, and the rails were only long wooden stringers with strap iron nailed on top. So undeveloped were the resources of the country that the builders of the Baltimore and Ohio Railroad in 1828 petitioned Congress to remit the duty on the iron which it was compelled to import from England. The trains consisted of a string of little cars, with the baggage piled on the roof, and when they reached a hill they sometimes had to be pulled up the inclined plane by a rope. Yet the traveling in these earliest days was probably more comfortable than in those which immediately followed the general adoption of locomotives. When, five or ten years later, the advantages of mechanical as opposed to animal traction caused engines to be introduced extensively, the passengers behind them rode through constant smoke and hot cinders that made railway travel an incessant torture.

Yet the railroad speedily demonstrated its practical value; many of the first lines were extremely profitable, and the hostility with which they had been first received soon changed to an enthusiasm which was just as unreasoning. The speculative craze which invariably follows a new discovery swept over the country in the thirties and the forties and manifested itself most unfortunately in the new Western States — Ohio, Indiana, Illinois, and Michigan. Here bonfires and public meetings whipped up the zeal; people believed that railroads would not only immediately open the wilderness and pay the interest on the bonds issued to construct them, but that they would become a source of revenue to sadly depleted state treasuries. Much has been heard of government ownership in recent years; yet it is nothing particularly new, for many of the early railroads in these new Western States were built as government enterprises, with results which were frequently disastrous. This mania, with the land speculation accompanying it, was largely responsible for the panic of 1837 and led to that repudiation of debts in certain States which for so many years gave American investments an evil reputation abroad.

In the more settled parts of the country, however, railroad building had comparatively a more solid foundation. Yet the railroad map of the forties indicates that railroad building in this early period was incoherent and haphazard. Practically everywhere the railroad was an individual enterprise; the builders had no further conception of it than as a line connecting two given points usually a short distance apart. The roads of those days began anywhere and ended almost anywhere. A few miles of iron rail connected Albany and Schenectady. There was a road from Hartford to New Haven, but there was none from New Haven to New York. A line connected Philadelphia with Columbia; Baltimore had a road to Washington; Charleston, South Carolina, had a similar contact with Hamburg in the same State. By 1842, New York State, from Albany to Buffalo, possessed several disconnected stretches of railroad. It was

not until 1836, when work was begun on the Erie
Railroad, that a plan was adopted for a single line
reaching several hundred miles from an obvious
point, such as New York, to an obvious destination, such as Lake Erie.   Even then a few farsighted men could foresee the day when the railroad train would cross the plains and the Rockies
and link the Atlantic and the Pacific.   Yet, in 1850
nearly all the railroads in the United States lay
east of the Mississippi River, and all of them, even
when they were physically mere extensions of one
another, were separately owned and separately
managed.

Successful as many of the railroads were, they
had hardly yet established themselves as the one
preëminent means of transportation.   The canal
had lost in the struggle for supremacy, but certain
of these constructed waterways, particularly the
Erie, were flourishing with little diminished vigor.
The river steamboat had enjoyed a development
in the first few decades of the nineteenth century
almost as great as that of the railroad itself.   The
Mississippi River was the great natural highway
for the products and the passenger traffic of the
South Central States; it had made New Orleans
one of the largest and most flourishing cities in the

country; and certainly the rich cotton planter of the fifties would have smiled at any suggestion that the "floating palaces" which plied this mighty stream would ever surrender their preëminence to the rusty and struggling railroads which wound along its banks.

This period, which may be taken as the first in American railroad development, ended about the middle of the century. It was an age of great progress but not of absolutely assured success. A few lines earned handsome profits, but in the main the railroad business was not favorably regarded and railroad investments everywhere were held in suspicion. The condition that prevailed in many railroads is illustrated by the fact that the directors of the Michigan and Southern, when they held their annual meeting in 1853, had to borrow chairs from an adjoining office as the sheriff had walked away with their own for debt. Even a railroad with such a territory as the Hudson River Valley, and extending from New York to Albany existed in a state of chronic dilapidation; and the New York and Harlem, which had an entrance into New York City as an asset of incalculable value, was looked upon merely as a vehicle for Wall Street speculation.

Meanwhile the increasing traffic in farm products, mules, and cattle from the Northwest to the plantations of the South created a demand for more ample transportation facilities. In the decade before the Civil War various north and south lines of railway were projected and some of these were assisted by grants of land from the Federal Government. The first of these, the Illinois Central, received a huge land-grant in 1850 and ultimately reached the Gulf at Mobile by connecting with the Mobile and Ohio Railroad which had also been assisted by Federal grants. But the panic of 1857, followed by the Civil War, halted all railroad enterprises. In the year 1856 some 3600 miles of railroad had been constructed; in 1865 only 700 were laid down. The Southern railroads were prostrated by the war and north and south lines lost all but local traffic.

After the war a brisk recovery began and brought to the fore the first of the great railroad magnates and the shrewdest business genius of the day, Cornelius Vanderbilt. Though he had spent his early life and had laid the basis of his fortune in steamboats, he was the first man to appreciate the fact that these two methods of transportation were about to change places — that water transportation was to decline and that rail transportation was to gain the ascendancy. It was about 1865 that Vanderbilt acted on this farsighted conviction, promptly sold out his steamboats for what they would bring, and began buying railroads despite the fact that his friends warned him that, in his old age, he was wrecking the fruits of a hard and thrifty life. But Vanderbilt perceived what most American business men of the time failed to see, that a change had come over the railroad situation as a result of the Civil War.

The time extending from 1860 to about 1875 marks the second stage in the railroad activity of the United States. The characteristic of this period is the development of the great trunk lines and the construction of a transcontinental route to the Pacific. The Civil War ended the supremacy of the Mississippi River as the great transportation route of the West. The fact that this river ran through hostile territory — Vicksburg did not fall until July 4, 1863 — forced the farmers of the West to find another outlet for their products. By this time the country from Chicago and St. Louis eastward to the Atlantic ports was fairly completely connected by railroads. The necessities of war led to great improvements in construction and equip-

constructed across the Hudson at this point. On the trains the little flickering oil lamps now gave place to gas, and the wood burning stoves — frequently in those primitive days smeared with tobacco juice — in a few years were displaced by the new method of heating by steam.

The accidents which had been almost the prevailing rule in the fifties and sixties were greatly reduced by the Westinghouse air-brake, invented in 1868, and the block signaling system, introduced somewhat later. In the ten years succeeding the Civil War, the physical appearance of the railroads entirely changed; new and larger locomotives were made, the freight cars, which during the period of the Civil War had a capacity of about eight tons, were now built to carry fifteen or twenty. The former little flimsy iron rails were taken up and were relaid with steel. In the early seventies when Cornelius Vanderbilt substituted steel for iron on the New York Central, he had to import the new material from England. In the Civil War period, practically all American railroads were single track lines — and this alone prevented any extensive traffic. Vanderbilt laid two tracks along the Hudson River from New York to Albany, and four from Albany to Buffalo, two exclusively for freight and two for passengers. By 1880 the American railroad, in all its essential details, had definitely arrived.

But in this same period even more sensational developments had taken place. Soon after 1865 the imagination of the American railroad builder began to reach far beyond the old horizon. Up to that time the Mississippi River had marked the Western railroad terminus. Now and then a road straggled beyond this barrier for a few miles into eastern Iowa and Missouri; but in the main the enormous territory reaching from the Mississippi to the Pacific Ocean was crossed only by the old trails. The one thing which perhaps did most to place the transcontinental road on a practical basis was the annexation of California in 1848; and the wild rush that took place on the discovery of the gold fields one year later had led Americans to realize that on the Pacific coast they had an empire which was great and incalculably rich but almost inaccessible. The loyalty of California to the Northern cause in the war naturally stimulated a desire for closer contact. In the ten years preceding 1860 the importance of a transcontinental line had constantly been brought to the attention of Congress and the project had caused much jealousy between

the North and the South, for each region desired
to control its Eastern terminus. This impediment
no longer stood in the way; early in his term, therefore, President Lincoln signed the bill authorizing
the construction of the Union Pacific — a name
doubly significant, as marking the union of the
East and the West and also recognizing the sentiment of loyalty or union that this great enterprise was intended to promote. The building of
this railroad, as well as that of the others which
ultimately made the Pacific and the Atlantic coast
near neighbors — the Santa Fé, the Southern Pacific, the Northern Pacific, and the Great Northern
— is described in the pages that follow. Here it is
sufficient to emphasize the fact that they achieved
the concluding triumph in what is certainly the
most extensive system of railroads in the world.
These transcontinental roads really completed the
work of Columbus. He sailed to discover the western route to Cathay and found that his path was
blocked by a mighty continent. But the first train
that crossed the plains and ascended the Rockies
and reached the Golden Gate assured thenceforth
a rapid and uninterrupted transit westward from
Europe to Asia.

CHAPTER II

A STORY was told many years ago of Commodore Vanderbilt which, while perhaps not strictly true, was pointed enough to warrant its constant repetition for more than two generations. Back in the sixties, when this grizzled railroad chieftain was the chief factor in the rapidly growing New York Central Railroad system, whose backbone then consisted of a continuous one-track line connecting Albany with the Great Lakes, the president of a small cross-country road approached him one day and requested an exchange of annual passes.

"Why, my dear sir," exclaimed the Commodore, "my railroad is more than three hundred miles long, while yours is only seventeen miles."

"That may all be so," replied the other, "but my railroad is just as wide as yours."

This statement was true. Practically no railroad, even as late as the sixties, was wider than

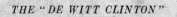

THE " DE WITT CLINTON "

The first locomotive and train in the State of New York.
Photograph from the original.

1.

another. They were all single-tracked lines. Even the New York Central system in 1866 was practically a single-track road; and the Commodore could not claim to any particular superiority over his neighbors and rivals in this particular. Instead of sneering at his "seventeen-mile" colleague, Vanderbilt might have remembered that his own fine system had grown up in less than two generations from a modest narrow-gage track running from "nothing to nowhere." The Vanderbilt lines, which today with their controlled and affiliated systems comprise more than 13,000 miles of railroad — a large portion of which is double-tracked, no mean amount being laid with third and fourth tracks — is the outgrowth of a little seventeen-mile line, first chartered in 1826, and finished for traffic in 1831. This little railroad was known as the Mohawk and Hudson, and it extended from Albany to Schenectady. It was the second continuous section of railroad line operated by steam in the United States, and on it the third locomotive built in America, the *De Witt Clinton*, made a satisfactory trial trip in August, 1831.

The success of this experiment created a sensation far and wide and led to rapid railroad building in other parts of the country in the years immediately following.   The experiences of a participant in this trial trip are described about forty years later in a letter written by Judge J. L. Gillis of Philadelphia:

In the early part of the month of August of that year [1831], I left Philadelphia for Canandaigua, New York, traveling by stages and steamboats to Albany and stopping at the latter place. I learned that a locomotive had arrived there and that it would make its first trip over the road to Schenectady the next day. I concluded to lie over and gratify my curiosity with a first ride after a locomotive.

That locomotive, the train of cars, together with the incidents of the day, made a very vivid impression on my mind. I can now look back from one of Pullman's Palace cars, over a period of forty years, and see that train together with all the improvements that have been made in railroad travel since that time. . . . I am not machinist enough to give a description of the locomotive that drew us over the road that day, but I recollect distinctly the general make-up of the train. The train was composed of coach bodies, mostly from Thorpe and Sprague's stage coaches, placed upon trucks. The trucks were coupled together with chains, leaving from two to three feet slack, and when the locomotive started it took up the slack by jerks, with sufficient force to jerk the passengers who sat on seats across the tops of the coaches, out from under their hats, and in stopping, came together with such force as to send them flying from the seats.

They used dry pitch for fuel, and there being no smoke or spark catcher to the chimney or smoke-stack, a volume of black smoke, strongly impregnated with sparks, coals, and cinders, came pouring back the whole length of the train. Each of the tossed passengers who had an umbrella raised it as a protection against the smoke and fire. They were found to be but a momentary protection, for I think in the first mile the last umbrella went overboard, all having their covers burnt off from the frames, when a general *mêlée* took place among the deck passengers, each whipping his neighbor to put out the fire. They presented a very motley appearance on arriving at the first station. Then rails were secured and lashed between the trucks, taking the slack out of the coupling chains, thereby affording us a more steady run to the top of the inclined plane at Schenectady.

The incidents off the train were quite as striking as those on the train. A general notice of the contemplated trip had excited not only the curiosity of those living along the line of the road, but those living remote from it, causing a large collection of people at all the intersecting roads along the route. Everybody, together with his wife and all his children, came from a distance with all kinds of conveyances, being as ignorant of what was coming as their horses, and drove up to the road as near as they could get, only looking for the best position to get a view of the train. As it approached the horses took fright and wheeled, upsetting buggies, carriages, and wagons, and leaving for parts unknown to the passengers if not to their owners, and it is not now positively known if some of them have stopped yet.

Such is a hasty sketch of my recollection of my first ride after a locomotive.

The Mohawk and Hudson Railroad was originally constructed with inclined planes worked by stationary engines near each terminus, the inclinations being one foot in eighteen. The rail used was a flat bar laid upon longitudinal sills. This type of rail came into general use at this period and continued in use in parts of the country even as late as the Civil War.

The roads that now make up the New York Central were built piecemeal from 1831 to 1853; and the organization of this company in the latter year, to consolidate eleven independent roads extending from Albany to Buffalo, finally put an end to the long debate between canals and railroads. The founding of this company definitely meant that transportation in the United States henceforth would follow the steel route and not the water ditch and the towpath. Canals might indeed linger for a time as feeders, even, as in the case of the Erie and a few others, as more or less important transportation routes, but every one now realized that the railroad was to be the great agency which would give plausibility to the industrial organization of the United States and develop its great territory.

Besides the pioneer Mohawk and Hudson, this consolidation included the Utica and Schenectady, which had been opened in 1836 and which had operated profitably for many years, always paying large dividends. The Tonawanda Railroad, opened in 1837, and the Buffalo and Niagara Falls, also finished in the same year, were operated with profit until they were absorbed by the new system. In 1838 the Auburn and Syracuse and the Hudson and Berkshire Railroads were opened. The former after being merged in 1850 with the Rochester and Syracuse Railway, became a part of the consolidation. The Syracuse and Attica Railroad, opened in 1839, the Attica and Buffalo, opened in 1842, the Schenectady and Troy, opened in the same year, and several other small lines, some of which had undergone various changes in name and ownership, were all merged into the New York Central Railroad. This great property now comprised five hundred and sixty miles of railroad, the main stem extending from Albany to Buffalo. Though it had as yet no connection with the Hudson River Railroad, the New York Central Railroad at this period was the most substantial and important of American railroad systems. It developed a large and healthy through traffic to the Great Lakes and was practically free from railroad competition. The Erie Railway, which for many years had been struggling under great difficulties to reach the Great Lakes and had gone through nearly a generation of financial vicissitudes, was just getting its through line actively under way. The Pennsylvania Railroad was just pushing through to the waters of the Ohio and was not likely for many years to compete with the New York Central for the lake traffic. The Baltimore and Ohio, while remotely a competitor, was, like the Pennsylvania, looking more for the traffic of the Ohio Valley than for that of the Lakes.

The period of six years following the consolidation of 1853 was one of great prosperity for the New York Central system, and, notwithstanding the setbacks to business caused by the panic of 1857, large dividends were continuously paid on the capital stock. In the year 1859 — before the Vanderbilt régime opened — the management embraced what to modern men of affairs are famous names. Erastus Corning was president, Dean Richmond was vice-president, and John V. L. Pruyn, Nathaniel Thayer, Isaac Townsend, and Chauncey Vibbard were directors. The headquarters of the company were at Albany, and the stock was owned mainly by residents of that city.

Meanwhile the building of railroads in other parts of the State and under other leadership was going forward rapidly. As far back as 1832 the first mile of the New York and Harlem Railroad was opened for traffic. This single mile remained for some time the only property of the company. It extended through what is now a thriving part of down-town New York. Its original terminus was at Prince Street, but the line was afterwards extended southward to the City Hall and later to the Astor House. It was not until 1837 that the road reached northward to Harlem and not until 1842 that Williamsbridge became the northern terminus. The line was looked upon as a worthless piece of property until 1852, when it was extended north to Chatham, to connect with the Albany and Stockbridge Railroad, and thus give a through line from New York City to Albany.

Another property built in these days and destined to become eventually an important part of the Vanderbilt lines was the Hudson River Railroad. This company was chartered in 1846, but for many years was frowned on as an unsound business venture, because of the belief that it would be in direct competition with the river traffic and therefore could never be made to pay.

Nevertheless the promoters went ahead and by 1850 the road had been opened to Poughkeepsie. The entire line of one hundred and forty-four miles was completed to East Albany in 1851. At the same time the Troy and Greenbush Railroad, extending six miles to Troy, was leased, thus giving the new Hudson River Railroad an entry into the city of Troy. The Hudson River Railroad was entirely independent of the New York Central enterprise and was controlled in those early days by a group of New Yorkers, prominent among whom was Samuel Sloan.

As we enter the Civil War period, we find the three important properties which were afterwards to make up the Vanderbilt system all developing rapidly and logically into the strategical relationship which would make ultimate consolidation inevitable. The completion of the Erie Railway and its gradual development as the only through line across the State from New York to the Great Lakes; the opening, expansion, and general solidification of the Pennsylvania lines and their aggressive policy of reaching out to the lake region on the west and across New Jersey on the east; the extension of the Erie interests into the New England field, and the possibility that the latter might gain control of the Harlem or the Hudson River Railroad — all these considerations naturally aroused in the New York Central interests a desire to insure the future by obtaining for themselves control of the lines that would connect their own system with New York City and the Eastern seaboard.

During the Civil War, however, no progress was made in this direction. It was not until 1869, four years after the closing of the war, that any radical change took place. But in the years that had intervened, a new and commanding figure in the railroad world had come upon the scene. This man had grown to be the dominating genius, not only in the field of railway expansion, but in the world of finance as well. His name was Cornelius Vanderbilt. Born in 1794 in very humble circumstances, he had received little or no education, and as a youth had eked out a living by ferrying passengers and garden produce from Staten Island to New York. He had painfully saved a few hundred dollars within a year or two after his marriage, and with this capital he began his career in the transportation business. From his first ferrying project he engaged in other undertakings and laid the foundation of his subsequent fortune in steamboat navigation. About 1860, at an age when most and the Hudson River Railroad. To get the latter two roads under his complete control was Vanderbilt's first object. He would then have unimpeded access to New York and so become independent of the river.

He began his ambitious plans by making himself the master of the Harlem property, and in so doing got his first experience in railroad stock manipulation and at the same time picked up a moderate fortune. It was comparatively easy to buy the control of the Harlem Railroad. The Company had never paid a dividend, and, in 1863, when the Commodore quietly began his work, the stock was selling below thirty dollars a share. Before the close of this year he had manipulated the stock until it had reached ninety-two, and by a corner, in August of that year, he raised it to 179. On this deal Vanderbilt reaped a nice little fortune — but evidently not enough to enable him to carry through the ambitious plans which were in the back of his head, for in 1864 we find him manipulating another corner and this time running the price of the stock up to 285. In this wise the Commodore not only added millions to his already growing fortune but also made himself a power in the financial world. Financiers began to fear him.

and he found it comparatively easy later to buy up the control of the Hudson River Railroad, which he did by paying about 100 for the stock. Then he began speculating again, sent Hudson River up to 180, and incidentally reaped another fortune for himself.

By this time Vanderbilt had achieved a great reputation as a man who created values, earned dividends, and invented wealth as if by magic; other railroad managers now began to lay their properties at his feet and ask him to do with them what he had done with the Harlem and the Hudson River. For under the Commodore's magic touch the Harlem Railroad for the first time in its long history began to pay dividends at a high rate, and in four years the earnings of the Hudson River property had nearly doubled.

One of the first properties to be placed at Vanderbilt's feet was the New York Central, and the control passed into his hands in the winter of 1866–67. He was now in a powerful position and immediately began to lay his plans for obtaining control of the Erie Railroad in the following year. In the latter effort he did not succeed, however, and after a protracted and dramatic contest he was defeated by his great adversary, "Uncle"

Daniel Drew. The story of this contest need not be detailed here, as it is given in full in the chapter on the Erie Railroad.

In the fall of 1869 the Commodore, having secured everything in the railroad field he had sought except the Erie, put through his scheme for consolidation. The New York Central and Hudson River Railroad was incorporated. It included the old New York Central and also the Hudson River Railroad but not the Harlem. The capital of the consolidated company was placed at ninety million dollars, a figure of such magnitude in those days that the world was startled. The system embraced in all nearly 850 miles of railroad lines. A few years later the Harlem Railroad was leased to the property at a high valuation and a large dividend was guaranteed on the stock, the ownership of which was retained by the Vanderbilt family.

The Vanderbilt system as it is now understood really began with these transactions. From this time on, its history has been similar in many respects to that of other large systems which were the outgrowth of merger or manipulation in these early days. During the remarkable period of commercial and industrial development in this country from 1870 onward, when thousands of miles of

3

new lines were built every year, when the growth of population was beginning to make the States of Ohio, Indiana, and Illinois centers of wealth and production, and when the wonderful Northwestern country embracing the States of Michigan, Wisconsin, and Minnesota, was so rapidly opened up and brought nearer to the Eastern markets, the Vanderbilt railroad interests were not idle. The original genius, Cornelius Vanderbilt, was soon gathered to his fathers, but his son, William H. Vanderbilt, was in many ways a worthy successor.

By 1885 the Vanderbilt lines had grown in extent and importance far beyond any point of which the elder Vanderbilt had ever dreamed. Long before this year the system included many smaller lines within the State of New York, and it had also acquired close control of the great Lake Shore and Michigan Southern system, with its splendid line from Buffalo to Chicago, consisting of more than 500 miles of railroad; the Michigan Central, owning lines from Detroit to Chicago, with many branches in Michigan and Illinois; the Canada Southern Railway, extending from Detroit to Toronto; and in addition to all these about 800 miles of other lines in the States of Ohio, Indiana, Michigan, and Pennsylvania.

been merged in the section extending westward from Chicago to Omaha and radiating throughout Iowa, Minnesota, Kansas, Wisconsin, Missouri, and other States. This company was known as the Chicago and North Western Railroad, and its property, which was one of large and growing value, by 1886 embraced a system of over 3500 miles of road. Although neither controlled by the New York Central nor directly affiliated therewith, it was classed as a Vanderbilt property.

While for many years after the death of the Commodore the Vanderbilt family remained in direct financial and operating control of the New York Central and its myriad of subsidiary lines and their genius as railroad builders and operators was distinctly evident, yet the brains and resources of the Vanderbilts were not alone responsible for the brilliant career of the system down to recent times. William H. Vanderbilt, though a man of unusual ability, did not possess the breadth of view or the sagacity of his father, and in the course of a few years he found himself exposed to a cyclone of public criticism. He had let it be widely known that he was personally the owner of over eighty-seven per cent of the hundred million capital of the company. In 1879 the New York Legislature, backed by the force of the popular anger and surprise at the accumulation of a hundred million dollar fortune by one man in ten years, was investigating the management of the New York Central with a view to curtailing its power; the rate wars were on between the seaboard and Chicago; and Jay Gould was threatening to divert all the traffic of his Wabash, St. Louis, and Pacific lines from the New York Central and turn it over to other Eastern connections unless Vanderbilt would give him a vital interest in the Vanderbilt lines.

Vanderbilt was harassed beyond endurance and, being of softer material than his father, was fearful of the outcome of public opinion, notwithstanding the fact that in a moment of anger — according to the statement of a newspaper reporter whose veracity Vanderbilt denied to his dying day — he had used the familiar expression, "The public be damned!" There were intimations that the Legislature was planning to impose heavy taxes on the property, solely because Vanderbilt held this gigantic personal ownership in the property. This prospect frightened him and he consulted friends whose judgment he respected. They urged him to sell a considerable part of his holdings in order to distribute the ownership of the property among a large number of people.

This plan could not be carried out, however, in the ordinary way, because large sales of stock by the Vanderbilt interests, if the speculating and investing public learned that he was making them, would greatly depreciate the price and might create general demoralization and a panic, while they would certainly injure the credit of the New York Central property. But a way out of the dilemma had to be found. It was at this juncture that a new personality, later to be closely identified with the Vanderbilt lines for a long series of years, appeared upon the scene. Vanderbilt was advised to consult J. Pierpont Morgan, of the banking house of Drexel, Morgan and Co. At that time the name of J. P. Morgan was just beginning to come prominently to the front in banking circles in New York. The Drexels had been conspicuous in business in Philadelphia for many years and in a sense were the fiscal agents of the great Pennsylvania Railroad Company. But the spectacular success of the House of Morgan a few years before in marketing the French government loan in England had added largely to its prestige. And so Vanderbilt concluded that, if any man could show him a way out in his difficult problem, Pierpont Morgan was that man.

The upshot of the matter was that Morgan devised a plan for the sale of a large amount of Vanderbilt's stock holdings through private sale in England, and in such a way that the knowledge of such sale would not become public in America. A confidential syndicate was formed which undertook to take the stock in a block and pass it on to English investors at approximately its current market price of about $130 per share. The sale was promptly accomplished; the stock went into the hands of unknown interests abroad; Vanderbilt received more than $25,000,000 in cash, which he largely reinvested in United States government bonds, and the Morgan syndicate reaped a profit of about $3,000,000. Five months after the closing of the syndicate public announcement was made of the sale and of the syndicate profit. The striking success of this transaction naturally added greatly to the prestige of J. P. Morgan as a financier of very large caliber, and it had the satisfactory effect of curtailing the legislative attacks on Vanderbilt.

From that date forward, the history of the Vanderbilt railroads has been closely identified with the House of Morgan. J. P. Morgan and his business associates became the company's financial agents, and thereafter all plans of expansion or consolidation were handled directly by them. In the board of directors Morgan banking interests had full representation, which they have held until this day.

The subsequent history of the Vanderbilt lines is chiefly a story of business expansion and growth. From 1885 to 1893, the great panic year, the New York Central each year added to its mileage, either by merger of smaller lines or by construction. All this time it was consolidating the system, eliminating the weaker links, and strengthening the stronger. Its lines penetrated all the best Eastern railroad territory outside of New England, New Jersey, and Pennsylvania, and no other railroad system in the country, with the single exception of the Pennsylvania, covered anything like the same amount of rich and settled territory, or reached so many cities and towns of importance. New York, Buffalo, Cleveland, Detroit, Chicago, St. Louis, Cincinnati, Indianapolis — these are a few of the great traffic centers which were included in the Vanderbilt preserves. The population of all these cities, as well as that of the hundreds of smaller

places and the countryside in general, was growing
by leaps and bounds. Furthermore the Northwest, beyond the Great Lakes and through to the
Pacific coast, saw the beginnings of its great development at this time; and the wheat fields of the
far western country became a factor of profound
importance in the national development. Consequently when the period of depression arrived with
the panic of 1893, the Vanderbilt properties were,
as a whole, in a strong position to meet the changed
situation and, like the great Pennsylvania property, they all passed through to the advent of the
new industrial era without the defaulting of a bond
or the passing of a dividend. The remarkable character of this achievement is evident in view of the
fact that in the period from 1893 to 1898 more than
sixty-five per cent of all the railroad mileage in the
United States went into the hands of receivers.

After the close of this era of panic, the Vanderbilt lines began expanding again, though on a
much smaller scale than in their more active time.
In 1898 William K. Vanderbilt, then president,
made the announcement that the New York Central had leased the Boston and Albany Railroad,
at that time a lucrative line running from Albany
across Massachusetts into Boston. This gave the

system an entry into the New England field, which it has continuously held since. A few years later this New England interest was increased by the acquisition of the Rutland Railroad in Vermont, thus making connection with the Ogdensburg and Lake Champlain, a line running across the northern part of New York State, which had also come under Vanderbilt control.

When business revived in the closing years of the nineteenth century, the history of American railroads began a new chapter. Federal railroad regulation, which started in a moderate way with the passage of the Interstate Commerce Act in 1887, had steadily increased through the years; the Sherman Anti-trust Act, passed in 1890, had been interpreted broadly as affecting the railroads of the country as well as the industrial and other combinations. These influences had thus greatly curtailed the consolidation of competing lines which had gone on so rapidly during the decades following the Civil War. Railroad managers and financiers therefore began to face a very serious problem. Competition of a more or less serious nature was still rampant, rates were cut, and traffic was pretty freely diverted by dubious means. Consequently many large railroad systems of heavy capitalization bid fair to run into difficulties on the first serious falling off in general business.

Great men are usually the products of their times and one of the men developed by these times takes rank with the greatest railroad leaders in history. Edward H. Harriman had risen in ten years from comparative obscurity and was now the president of the Union Pacific Railroad, which he had, in conjunction with the banking house of Kuhn, Loeb and Company, reorganized and taken out of bankruptcy. Harriman was one of the originators of the "community of interest" idea, a device for the partial control of one railroad system by another. For instance, although the law forbade any railroad system from acquiring a complete control of a competing line by purchasing a majority of its capital stock or by leasing it, nothing was said about one railroad having a minority investment interest in another. A minority investment, even though it be as low as ten or twenty per cent, usually constitutes a dominating influence if held by a single interest, for in most cases the majority of the shares will be owned in small blocks by thousands of investors who never combine for a definite, practical purpose. Thus the interest which has the one large block of stock usually controls the voting power, and runs little risk of losing it unless a contest develops with other powerful interests — and this is a contingency which it almost never has to meet.

Carrying out this policy of promoting harmony among competing lines, the New York Central and Pennsylvania Railroad early in 1900 acquired a working control of the Reading Company, which in turn controlled the New Jersey Central and dominated the anthracite coal traffic. Later the Baltimore and Ohio shared this Reading interest with the Lake Shore of the New York Central system. The New York Central and the Pennsylvania acquired a working control of the same kind in the Chesapeake and Ohio Railway, which was an important element in the soft coal fields and was reaching out to grasp soft coal properties in Ohio and Indiana.

These and other purchases, and the consequent voice acquired in the management, established comparative harmony among Eastern railroads for a long time; they stabilized rates and enabled formerly competing roads to parcel out territory equitably among the different interests. Later, Harriman, and to some extent Morgan, carried the community of interest idea some steps further.

Morgan caused the New York Central to acquire stock interests in certain "feeder" lines such as the New York, New Haven and Hartford and the Chicago, Milwaukee and St. Paul, as well as in competing lines; and Harriman caused the Union Pacific not only to dominate the Southern Pacific Company by minority control but also to acquire interests in the Illinois Central, the Baltimore and Ohio, the New York Central, and other eastern properties. The fact was that Harriman had plans in view for acquiring actual control of the New York Central for the Union Pacific and thus, with the Illinois Central, of creating a continuous transcontinental line from ocean to ocean.

In the past decade few unusual or startling events have marked the history of the Vanderbilt lines. The Vanderbilt family no longer possesses a majority interest in the stock, or anything which approaches it, and the New York Central system and its subsidiaries have come to be known more and more as Morgan properties. The system has grown up with the country. Many of its former controlled roads have now been merged into the main corporation and many new lines have been added to it. Hundreds of millions of dollars of new capital have been spent on the main lines and

CHAPTER III

In the early forties the commercial importance of Philadelphia was menaced from two directions. A steadily increasing volume of trade was passing through the Erie Canal from the Central West to the northern seaboard, while traffic over the new Baltimore and Ohio Railroad promised a great commercial future to the rival city of Baltimore. With commendable enterprise the Baltimore and Ohio Company was even then reaching out for connections with Pittsburgh in the hope of diverting western trade from eastern Pennsylvania. Moreover the financial prestige of Philadelphia had suffered from recent events. The panic of 1837, the contest of the United States Bank with President Jackson, its defeat, and its subsequent failure as a state bank, the consequent distress in local financial circles — all conspired to shift the monetary center of the country to New York.

It was at this time that Philadelphia capitalists began to bestir themselves in an attempt to recover their lost opportunities. Philadelphia must share in this trade with the Central West. The designs of the Baltimore and Ohio Company must be defeated by bringing Pittsburgh into contact with its natural Eastern market. To this end, the Pennsylvania Railroad was incorporated on April 13, 1846, with a franchise permitting the construction of a railroad across the State from Harrisburg to Pittsburgh. An added incentive to constructive expansion was given by an act of the Legislature authorizing the Baltimore and Ohio to extend its line to Pittsburgh if the Pennsylvania Company failed to avail itself of its franchise.

In order to avoid the heavy cost of constructing a road between Philadelphia and Harrisburg, the Pennsylvania Railroad entered into arrangements with the Philadelphia and Columbia — a railroad opened in 1834 and owned by the State — which ran through Chester and Lancaster to Columbia. This road was primitive in the extreme and used both steam and horse power. As late as 1842 a train was started only when sufficient traffic was waiting along the road to warrant the use of the engine. Belated trains were hunted up by horsemen. Yet the road was in those days famous for the "rapidity and exceptional comforts of the train service." Between Columbia and Harrisburg passengers westward bound had to use the Pennsylvania Canal.

Construction of the main line westward to Pittsburgh began at once and progressed rapidly. By making use of the Alleghany Portage Railroad from Hollidaysburg, the Pennsylvania Railroad eventually secured a continuous line from Harrisburg to Pittsburgh. But between Philadelphia and Harrisburg passengers were for a long time subjected to many inconveniences. Finally in 1857 the Pennsylvania Railroad bought the Philadelphia and Columbia from the State, rebuilt it, and extended it to Harrisburg. At the same time the Pennsylvania bought the main line of the Public Works, which included the Alleghany Portage Railroad. On July 18, 1858, the first through train passed over the entire line from Philadelphia via Mount Joy to Pittsburgh without transfer of passengers. At the same time the first smoking car ever attached to a passenger train was used, and sleeping cars also soon began to appear.

The railroad genius identified with the history of the Pennsylvania Railroad during the following

4

decade is J. Edgar Thomson. A man of vision and of great shrewdness and ability, he was more like the modern railroad head of the Ripley or Underwood type than of the Vanderbilt, Garrett, or Drew type. His interest was never in the stock market nor in the speculative side of railroading but was concentrated entirely on the development and operation of the Pennsylvania Railroad system. His dreams were not of millions quickly made nor of railroad dominance simply for the power that it gave; his mind was concentrated on the growth and prosperity of a vast railroad system which would increase with the years, become lucrative in its operations, and not only radiate throughout the State of Pennsylvania but extend far beyond into the growing West.

Under the Thomson management, which lasted until 1874, the record of the Pennsylvania Railroad was one of progress in every sense of the word. While Daniel Drew was lining his pockets with loot from the Erie Railroad and Commodore Vanderbilt was piling up his colossal fortune through consolidation and manipulation, J. Edgar Thomson was steadily building up the greatest business organization on the continent. In 1860, the entire Pennsylvania Railroad system was represented merely by the main line from Philadelphia to Pittsburgh, with a few short branches. By 1869 the road had expanded within Pennsylvania alone to nearly one thousand miles and also controlled lines northward to the shores of Lake Erie, through the State of New York.

But the master accomplishment of the Thomson administration was the acquisition of the Pittsburgh, Fort Wayne and Chicago line in 1869. This new addition gave the Company its own connection with Chicago and made a continuous system from the banks of the Delaware at Philadelphia to the shores of Lake Michigan, thus rivaling the far-flung Vanderbilt line, a thousand miles long, which the industrious Commodore was now organizing. Shortly thereafter the Pennsylvania began to expand on the east also and obtained an entry into New York City by acquiring the United Railroad and Canal Company, which owned lines across the State of New Jersey, passing through Trenton.

In the latter years of the Thomson management it became more and more evident that it was important for the Pennsylvania Railroad to have further Western connections which would reach the growing cities of the Middle West. While the

Fort Wayne route made a very direct connection with Chicago and included branches of value, yet the keen competition which was developing in the expansive years following the Civil War made actual control of the Middle Western territory a matter of sound business policy. The Vanderbilt lines were reaching out through Ohio, Indiana, and Illinois; the Baltimore and Ohio was steadily developing its Western connections, and now Jay Gould had come actively on the scene with large projects for the Erie. To offset these projects, early in 1870 a "holding company" — probably the first of its kind on record — known as the Pennsylvania Company was formed for the express purpose of controlling and managing, in the interest of the Pennsylvania Railroad, all lines leased or controlled or in the future to be acquired by the Pennsylvania Railroad interests west of Pittsburgh and Erie. This Company took over the lease of the Fort Wayne route and also acquired control by lease of the Erie and Pittsburgh, a road extending northward through Ohio to Lake Erie.

After this date the expansion of the system west of Pittsburgh went on rapidly. In 1871 the Cleveland and Pittsburgh Railroad, which had been opened as early as 1852, came under the Pennsylvania control. Soon after this, many smaller lines in Ohio were merged in the system. The most important acquisition during this period, however, was the result of the purchase of the great lines extending westward from Pittsburgh to St. Louis, with branches reaching southward to Cincinnati and northward to Chicago. This system — then known as the "Pan Handle" route and later as the Pittsburgh, Cincinnati, Chicago and St. Louis — was a consolidation of several independent properties of importance which had been gradually extending themselves over this territory during the previous decade. This new system, which embraced over fourteen hundred miles of road, gave the Pennsylvania a second line to Chicago, a direct line to St. Louis, a second line to Cincinnati, and access to territory not previously tapped.

While the achievements of the Pennsylvania Railroad Company during these years of consolidation and expansion are not to be compared with those of more modern times, it is well to realize that even as early as the seventh decade of the last century this railroad was always in the forefront in matters of high standards and progressive practice. It was the pioneer in most of the improvements which were later adopted by other roads. The Pennsylvania was the first American railroad to lay steel rails and the first to lay Bessemer rails; it was the first to put the steel fire-box under the locomotive boiler; it was the first to use the air brake and the block signal system; it was the first to use in its shops the overhead crane.

In these earlier years also the Pennsylvania had established its enviable record for conservative and non-speculative management. No railroad wrecker or stock speculator had ever had anything to do with the financial control of the company, and this tradition has been passed on from decade to decade. The stockholders themselves, even in those days of loose methods and careless finance, had the dominating voice in the affairs of the company and were also factors in the approval or disapproval of any proposed policies. In the matter of its finances the Pennsylvania developed and established an equally clean record. The company began almost at the beginning to pay a satisfactory dividend on its shares and continued to do so right through the Civil War period. Since the through line from Philadelphia to Pittsburgh was opened, not a single year has passed without the payment of a dividend — a sixty-year record which can be duplicated by no other American railroad system.

The Pennsylvania still continued to forge ahead even during the exciting period from 1877 to about 1889, when the trunk lines were aggressively carrying on that policy of cutthroat competition between Chicago and the Atlantic seaboard which resulted in so severely weakening the credit and position of properties like the Baltimore and Ohio and the Erie. The Pennsylvania, too, indulged in rate cutting, but the management was equal to the situation and made up in other directions what it lost in lower rates. It gave superior service, developed a high efficiency of operation, and steadily maintained its properties at a high standard. During these years the president was George B. Roberts, who had succeeded Thomas A. Scott in 1880.

Roberts's management spanned the period from 1880 to 1897 and embraced a decade of comparative prosperity for the country as a whole and nearly a decade of panic and industrial and financial depression. During the earlier decade the business of the Pennsylvania was continually benefited by the industrial development and growth which marked the period. It was at this time that

the Pittsburgh district took its permanent place
as the great center of steel and iron manufacture.
The discovery of petroleum in western Pennsylvania, creating an enormous new industry in itself,
proved to be an event of far-reaching significance
for the Pennsylvania Railroad. The extensive opening up of the soft coal sections of western Pennsylvania, Ohio, and Indiana, also meant much for this
great system of railroads.

Still further developments in other directions accrued to the benefit of the Pennsylvania Railroad.
In this period, by obtaining the control of a line
to Washington, the system acquired a Southern
artery running through Wilmington, Delaware,
and Baltimore to Washington. Afterwards, with
other roads, the Pennsylvania acquired control
of the Richmond, Fredericksburg and Potomac
Railroad and thus obtained a line to Richmond,
Virginia. On the north and to the east the expanding movement also went on. In addition to the
development of its main line from Philadelphia
to Jersey City, the Pennsylvania acquired many
other New Jersey lines, including the West Jersey
and Seashore, a road running from Camden to
Atlantic City and Cape May.

During the whole of the aggressive administration

of both Thomas A. Scott and George B. Roberts
the great system continued to spread out steadily
until it had penetrated as far as Mackinaw City
on the north and Chesapeake Bay on the south.
Its network of lines stretched across the Eastern
section of the continent from New York to Iowa
and Missouri, while the intensive development of
shorter lines in the State of Pennsylvania and to
the north was unceasing. The Northern Central
running south from Sodus Bay on Lake Ontario
through central Pennsylvania to Baltimore, the
Buffalo and Allegheny Valley extending from Oil
City northward and joining the main system to the
east, the Western New York and Pennsylvania operating north from Oil City to Buffalo and Rochester
— these lines the Pennsylvania Railroad acquired
and definitely consolidated in the Roberts régime.

After the retirement of Roberts, Frank Thomson, who had formerly been general manager,
was placed at the head of the system for three
years. But in 1899 Alexander J. Cassatt, who had
for many years been identified with the Pennsylvania as officer, director, and stockholder, took the
helm, and a new chapter and probably the greatest
in the history of this remarkable railroad began.

The name of Alexander J. Cassatt will always

be linked with the comprehensive terminal developments in the region of New York City which were begun almost immediately on his accession to the presidency and which were carried forward on bold and far-reaching lines. Perhaps more than any other one person, Cassatt foresaw the approach of the day when New York City as a commercial center would outstrip both in density of population and in amount of wealth all the other cities of the world. He and his predecessors had for many years witnessed the great industrial development of the Pittsburgh district, where property values had grown by leaps and bounds and where the steadily advancing development of industry and material resources had been so unmistakably reflected in the increasing earning power and value of the Pennsylvania Railroad properties.

But while at Pittsburgh the road had everything to favor it as far as terminals and rights of way through the heart of the great industrial district were concerned, in the great Eastern metropolis the Pennsylvania Railroad was at an obvious disadvantage, particularly as compared with the New York Central, which had its splendid terminal rights penetrating to the heart of the city. Cassatt saw that his company must without delay

take a number of bold and, for the time, enormously expensive steps toward the development
of terminal facilities in Greater New York or else
forever abandon the idea of getting nearer the
heart of the city than the New Jersey shore and
thus run the risk, in the keen contest for commercial supremacy, of ultimately falling behind other
more advantageously situated lines.

There were still further incentives to immediate
action on the part of the Pennsylvania Railroad.
While the New York Central was in an ideal position for handling all traffic destined for the New
England States, the Pennsylvania could control
practically none of this business, as its terminals
were on the wrong side of the Hudson and necessi-
tated not merely the inconvenient transfer of passengers but also the much more expensive handling of freight. Other disadvantages from which
the Pennsylvania suffered were involved in its
inability to make the most economical terms for
foreign shipping, as a large proportion of such
freight had to be constantly transferred on lighters
to the New York and Brooklyn sides of the harbor.
Thus any comprehensive plan for terminal development on the part of the Pennsylvania must
necessarily include not only a tunnel system into

New York City but also an outlet through the city to Long Island and a connection with the New England railroads.

The first move in the development of this terminal system was the acquisition in 1900 of the control of the Long Island Railroad, embracing all the steam railway mileage on Long Island, with lines extending along both the north and south shores to Montauk Point. This acquisition added extensive freight yards and terminals on the Brooklyn side of the East River. The Company then obtained franchises and began the construction of its great tunnels under the North and East Rivers and entirely across New York City, with a mammoth passenger station at Seventh Avenue and Thirty-second Street. A great railroad bridge was planned to cross from Long Island to the mainland, connecting with the New York, New Haven and Hartford system, in the stock of which the Pennsylvania at this time purchased an interest.

The terminal construction occupied a period of many years and cost over one hundred million dollars, besides the added costs involved in building up and developing the old, worn-out Long Island Railroad. Only recently has the project been rounded out and completed through the final construction of the important connection with the New England railroad systems. But the realization of this plan is undoubtedly the greatest achievement in all the long career of the Pennsylvania Railroad. Had the project been delayed for another decade, it probably could not have been accomplished because of the growing expense of operation and the difficulties of getting franchise rights and rights of way through and under the metropolis.

While the tunnel development is the notable achievement of the Cassatt régime, this remarkable man's name is also closely identified with the "community of interest" idea already explained. This "community of interest" scheme was pushed aggressively by Cassatt in coöperation with Harriman, Hill, and Morgan. Large stock purchases were made in the Norfolk and Western, the Chesapeake and Ohio, and the Baltimore and Ohio. As the latter road had in its turn acquired, jointly with New York Central interests, a working control of the Reading Company, and the Reading Company had secured majority ownership of the New Jersey Central system, it is apparent that the domination which the Pennsylvania had obtained over the entire Eastern seaboard south of New York City and north of Baltimore was made nearly complete.

The "community of interest" plan held sway with the large railroads of the country and was very effective for perhaps half a dozen years, until the interstate commerce laws were amended in such a way as to give the Government complete control over railroad freight and passenger rates. In 1906 the Pennsylvania began to dispose of the bulk of its holdings in competing properties, the most notable transactions being the sale of its entire interest in the Chesapeake and Ohio to independent interests and a substantial part of its Baltimore and Ohio holdings to the Union Pacific Railroad. A few years later, when the Union Pacific was forced by the Federal courts to dispose of its control of the Southern Pacific Company, a trade was made between the Pennsylvania and the Union Pacific whereby the latter took from the Pennsylvania the remainder of its Baltimore and Ohio investment and gave in exchange a portion of its own large holding of Southern Pacific stock.

To get a fair idea of the meaning and magnitude of the great Pennsylvania Railroad system today one must do more than scan maps and study statistics. One should travel by daylight over its main line from New York to Pittsburgh. Although the route is over the same ground which the road followed a generation or two ago, a four-track line runs practically all the way, with long stretches of hundreds of miles of five, six, and eight tracks. Where mountains were climbed thirty years ago, one will now find them bored by tunnels; where sharp curves were necessary before straight trackage only will be encountered today. Grades have been eliminated everywhere and the whole route has been modernized and strengthened by the laying of one hundred to one hundred and fifty pound rails.

Undoubtedly the fortunate location of the Pennsylvania lines in the half dozen States which represent the financial and industrial heart of the continent has had much to do with its vast growth and the expansion of its business; but its high reputation can be explained only by the long record of its superior methods and management. One of the primary objects of Pennsylvania Railroad policy has been to keep pace with the growth of the country. Instead of following in the wake of industrial progress and making its improvements and extensions after its competitors had made theirs, its management has usually had the foresight to prepare well in advance for future needs.

CHAPTER IV

THE ERIE RAILROAD

BEFORE introducing a friend to a distinguished stranger, it is advisable to give him some account of the person whose acquaintance he is about to make; and so, fellow-traveler, whom I introduce to the New York and Erie Railroad, it may be well to prefix here a brief sketch of the history and present condition of this, the Lion of Railways. True, he is yet in an unfinished state, but you will find that what there is of him is complete, and of wondrous organization and activity. His magnificent head and front repose in grandeur on the shores of the Hudson; his iron lungs puff vigorously among the Highland fastnesses of Rockland; his capacious maw fares sumptuously on the dairies of Orange and the game and cattle of Broome; his lumbar region is built upon the timber of Chemung, and the tuft of his royal extremity floats triumphantly on the waters of Lake Erie.

This exultant, characteristically American, description appeared in Harper's *Guide-Book of the New York and Erie Railroad*, published in 1851, soon after the opening of the main line of more than four hundred and sixty miles from Piermont on the Hudson to Dunkirk on Lake Erie. That this railroad, which after nearly twenty years of struggle and of financial vicissitudes had finally linked the Great Lakes with the Atlantic coast, was looked upon as a property of wonderful character and limitless future is indicated in all the railroad literature of that time. Appleton's *Illustrated Handbook of American Travel*, published in 1857, devotes several pages to a description of this remarkable achievement in railroad extension and among other things says:

This great route claims a special admiration for the grandeur of the enterprise which conceived and executed it, for the vast contribution it has made to the facilities of travel, and for the multiplied and various landscape beauties which it has made so readily and pleasantly accessible. It traverses the southern portion of the Empire State in its entire length from east to west, passing through countless towns and villages, over many rivers, through rugged mountain passes now, and anon amidst broad and fertile valleys and plains. In addition it has many branches, connecting its stations with other routes in all directions, and opening new stores of pictorial pleasures. . . . An interesting feature of this road is its own telegraph, which runs by the side of the road and has its operator in nearly every station house. This telegraph has a double wire, enabling the company to transact the public, as well as their own private business. Daily trains leave for the west on this route, with connections by boat from the foot of Duane Street, morning, noon, and night.

The Erie Railroad system was foreshadowed in the time of Queen Anne, when the Colony of New York appropriated the sum of five hundred dollars to John Smith and other persons for the purpose of constructing a public road connecting the port of New York with the West in the vicinity of the Great Lakes. The appropriation was coupled with the condition that within two years the beneficiaries should have constructed a road wide enough for two carriages to pass, from Nyack on the Hudson River to Sterling Iron Works, a distance of about thirty miles; and that they should cut away the limbs of trees over the track in order to allow the carriages to pass. In this way began the internal improvement system of the State of New York, which after the lapse of more than a century resulted in opening the Erie Canal and in projecting a railroad system connecting New York and the valley of the Hudson with Lake Erie.

After the opening of the Erie Canal in 1825, the Legislature of New York directed a survey of a state road which was to be constructed at public expense through the southern tier of counties from the Hudson River to Lake Erie. The unfavorable profile exhibited in the survey apparently caused the project to be abandoned. But the idea still held sway over the minds of many people; and the great benefits brought to the Mohawk Valley and surrounding country by the Erie Canal led the southern counties to demand a transportation route which would work similar wonders in that region. This growing sentiment finally persuaded the Legislature to charter in April, 1832, the New York and Erie Railroad Company, and to give it authority to construct a line and to regulate its own charges for transportation.

During the following summer a survey of the route was made by Colonel De Witt Clinton, Jr., and in 1834 a second survey was made of the whole of the proposed route. When the probable cost was estimated, many opponents arose who declaimed the undertaking was "chimerical, impractical, and useless." The road, they declared, could never be built and, if built, would never be used; the southern counties were mountainous, sterile, and worthless, and afforded no products requiring a market; and, in any case, these counties should find their natural outlet in the valley of the Mohawk. This antagonism was successfully opposed, however, and the construction of the road was begun in 1836.

The panic of 1837 interfered with the work, but in 1838 the state Legislature came forward with a construction loan of three million dollars, and the first section of line, extending from Piermont on the Hudson to Goshen, was put into operation in September, 1841. In the following year the company became financially embarrassed and was placed in the hands of receivers. This catastrophe delayed further progress for years, and it was not until 1846 that sufficient new capital was raised to go on with the work. The original estimate of the cost for building the entire line of 485 miles had been three million dollars, but already the road had cost over six millions and only a small portion had been finished. The final estimate now rose to fifteen millions, and, although some money was raised from time to time and new sections were built, there was no certainty that the entire road would ever be completed. Ultimately the State of New York canceled its claim against the property, new subscriptions of some millions were secured, and more money was raised by mortgaging the finished sections.

Finally, in 1851, after eighteen years of effort,

the line was opened to Lake Erie. In addition there had been added various feeders or branches, giving the road an entry into Scranton, Pennsylvania, and into Geneva and Buffalo, New York. It had its terminus on Lake Erie at Dunkirk and its eastern terminus at Piermont, near Nyack on the Hudson, about twenty-five miles by boat from New York City.

The financial condition of the Erie at this time manifested the beginning of that general policy of improvidence and recklessness which afterward, for nearly a generation and a half, made the company a speculative football in some of the most disreputable games of Wall Street stock-jobbers. For though the original estimate had been three millions and the highest estimate of the cost during construction had been fifteen million dollars, the company, in 1851, started its career with capital obligations of no less than twenty-six millions — a very large sum for those days.

The fact that these initial obligations constituted a heavy burden became apparent when the Erie began operations. They made necessary such high freight rates that shippers held indignation meetings and again and again made appeals for legislative relief. Although much money had been raised after 1849 for improvements, the condition of the Erie steadily grew worse. It soon became notorious for many accidents due to carelessness in running trains and to the breaking of the brittle iron rails.

But in spite of these drawbacks the business of the Erie grew. In 1852 it acquired the Ramapo and Paterson and the Paterson and Hudson River railroads and in this way it obtained a more direct connection with New York City. It changed the tracks of its new railroads to the six-foot gage, which the Erie had adopted from the start and which it persisted in maintaining for many years despite the world-wide practice of establishing a standard width of four feet eight and one-half inches.

The most conspicuous figure in the history of the Erie Railroad system in these early days was Daniel Drew. From 1851, when the main line was opened, until 1868, this man was a director and, for the larger part of the time, treasurer. Born in 1797, he had driven cattle when a boy from his native town of Carmel in Putnam County to the New York City market and, for some years later, he had been proprietor of the Bull's Head Tavern. Shrewd, unscrupulous, illiterate, good-natured, and manipulation. The stock of the Erie road was then selling at about 95 and the company was in pressing need of funds. The treasurer came to the rescue as usual and made the necessary advances on adequate security. The company had in its treasury a considerable amount of unissued stock and had also the legal right to issue bonds to the extent of $3,000,000 which could be converted into stock. Drew took these bonds and the unissued stock as security for a loan of $3,500,000.

It so happened, naturally, that Drew was soon heavily short of Erie stock in Wall Street. The market was buoyant; speculation was rampant and the outside public, the delight and prey of Wall Street gamblers, were as usual drawn in by the fascination of acquiring wealth without labor. At this time our friend, Daniel Drew, was quietly selling Erie stock and closing contracts for the future delivery of the certificates; and he was doing this at rising prices. As the days went by, his grave, desponding manner grew more and more apparent. Erie stock continued to rise. In the loan market its scarcity became greater hour by hour. The rumor began to spread that "Uncle Daniel" was cornered. His large obligations for future delivery must be met. Where was the Er steam railroads. When finally, in 1868, they crossed swords in connection with the two railroad systems extending through New York State, both were more than seventy years old and had been successful in the acquisition of millions by methods of their own invention. They were no doubt equally unscrupulous, but, while Drew was by nature a pessimist and "bearish," Vanderbilt, in the Wall Street vernacular, was always a "bull."

Having obtained control of the New York Central, the Hudson River, and the Harlem railroads, Commodore Vanderbilt now decided in the summer of 1867 to go after the Erie, of which Drew was nominally in possession, although no one knew when he owned a majority of the stock or when he was temporarily short of it. Usually he loaded up as the annual election of officers approached and liquidated shortly thereafter. Besides Vanderbilt there was another interest at this time trying for the control of the Erie. This interest consisted of certain Wall Street speculators and certain Boston capitalists who proclaimed themselves railroad reformers. These so-called reformers were as unscrupulous and crafty as either of the other men, and they really represented nothing but an attempt to raid the Erie treasury in the interest of a bankrupt New England corporation known as the Boston, Hartford and Erie Railroad. As was well said, the name of this latter road was "synonymous with bankruptcy, litigation, fraud, and failure."

The Erie Railroad control was always nominally for sale, and, as the annual election approached, a majority of stockholders stood ready to sell their votes to the highest bidder. Commodore Vanderbilt cleverly secured the coöperation of the "reformer" element, corralling their proxies, and thus he appeared to be in a position to oust the Drew interests without difficulty. On the Sunday preceding the election the Commodore saw Drew and amicably explained to him the trap he had laid, and showed him clearly that there was no way out of the situation. Upon this disclosure, Treasurer Drew at once faced about and agreed to join hands with Vanderbilt in giving the market for the stock the strong upward twist it had lacked before that hour. Jointly they would make so much money that neither side would lose anything. "Uncle Daniel" went away apparently satisfied and contented with the compromise.

But the Commodore had not finished. A few hours later he took the Boston adventurers into his confidence and explained that he proposed to continue Drew in the directorate. The Boston men were puzzled and confused by this sudden change of front. Later, all parties met at Drew's house, and Vanderbilt brought the Boston men to terms by proposing a plan to Drew whereby they would be entirely left out. This ruse succeeded and a written agreement to the advantage of all, but at the expense of the outside stockholders and of the general public, was then drawn up.

This, however, was only the beginning of the fight. Vanderbilt was now in the Erie as a joint owner, but he had stretched out his hands to control the road and he meant to succeed. In February of 1868, Frank Work, the single representative of Vanderbilt on the Erie board, applied for an injunction against Treasurer Drew and his brother directors to restrain them from the repayment of the $3,500,000 borrowed by the railroad from Drew in 1866, and to restrain Drew from taking any legal steps toward compelling a settlement. Judge Barnard granted a temporary injunction, and two days later Vanderbilt's attorney petitioned for the removal from office of Treasurer Drew. The papers presented in the case exposed a new fountain of Erie stock which had up to that time been entirely overlooked.

A recently enacted law of the State of New York — probably fathered by Drew — authorized any railroad company to create and issue its own stock in exchange for the stock of any other railroad under lease to it. Upon the basis of this law Drew and his close satellites had secretly secured ownership of a worthless piece of road connecting with the Erie and known as the Buffalo, Bradford and Pittsburgh. Then, as their personal needs in the stock-market or at elections demanded, they had supplied themselves with new Erie stock by leasing the worthless road to the Erie and then exchanging Erie stock for the worthless stock of the leased line. The cost of the line to Drew and his friends, as financiers, was about $250,000. They then issued, as proprietors, $2,000,000 in bonds of the road, payable to one of themselves as trustee. This person then shifted his character, became counsel for both sides, and drew up a contract leasing the line to the Erie for 499 years, the Erie agreeing to guarantee the bonds in consideration. These men then reappeared as directors of the Erie and ratified the lease. After that it was a simple matter to divide the loot. The Erie was thus saddled with a $2,000,000 mortgage at seven per cent in addition to a further issue of capital stock.

Following the first injunction another was now issued restraining Drew and the Erie board from making any further issues of stock, by conversion of bonds or otherwise, and also forbidding the Erie to guarantee any further issues of bonds. An additional injunction forbade Drew from having any transactions in Erie stock or fulfilling any contracts until he had returned to the treasury the shares involved in his loan transaction of 1866 and in the purchase of the worthless Buffalo, Bradford and Pittsburgh road.

Matters now looked forbidding for Treasurer Drew. Instead of being allowed to manufacture fresh Erie stock certificates at his own will, as had been his habit for fifteen years, he was to be cornered by a legal writ and forced to work his own ruin. But notwithstanding the apparently desperate situation it was quite evident that Drew's nerves were not seriously affected. Although he seemed rushing on destruction, he continued day after day to put out more short stock, all in the face of a steadily rising market. His plans, apparently, were carefully matured, and he said that if the Commodore wanted the stock of his road he would let him have all he desired — at the proper price.

As usual the Erie treasury was short of funds, and as usual "Uncle Daniel" stood ready to advance all the money required — that is, on proper security. There was but one kind of security to offer and that was convertible bonds. No stock could be issued by the company for less than par, but convertible bonds could be disposed of by the directors at any price. A secret meeting of the executive committee was held, at which it was voted to issue immediately and to offer for sale $10,000,000 in convertible bonds at 72½. Drew's broker at once became the purchaser of $5,000,000 worth. In ten minutes after the meeting had adjourned, the bonds had been issued, their conversion into stock demanded and made, and certificates for 50,000 shares of stock deposited in a broker's safe, subject to Drew's order.

A few days later came the injunction, and Erie stock began to soar in the markets, in response to which "Uncle Daniel," who had been industriously selling his many thousands of new shares, now determined to bring up the reserves and let the eager buyers have the other five millions; but before the bonds could be converted the injunction had been served. The date for the return of the writ was Tuesday the 10th of March; but the Erie ring needed less time than this and decided on Monday the 9th as the day to defeat the corner.

Saturday and Sunday were busy days for Drew and his friends. All his brokers had been enjoined, so a dummy was made the nominal purchaser of the bonds. This dummy then made his formal demand for the conversion of the bonds and was refused. All this was done upon affidavit, as it was the plan of Drew to get from some judge a writ of mandamus to compel the Erie Railroad to convert the bonds. The stock certificates for which they were to be exchanged were signed in blank and made ready for delivery.

Drew had agreed to sell 50,000 shares of stock at 80 to the firms of which Jay Gould and James Fisk, Jr., were members; they were also Erie directors. On Monday morning, the 9th of March, the certificates, filled out in the names of these firms, were handed by the secretary to an employee who was directed to carry them from the office of the company in West Street to the transfer clerk in Pine Street. The messenger left, but in a moment or two returned to report to the apparently amazed secretary that Fisk had met him outside the door, had taken the certificates away from him, and "had run away with them." It was true. The stock certificates had been "stolen" and were beyond the control of an injunction. The stock certificates next appeared in every part of Wall Street.

On the same day the Erie representatives applied to Judge Gilbert of Brooklyn for an injunction on the ground that certain persons, including Judge Barnard, had entered into a conspiracy to speculate in Erie stock and to use the process of the courts to aid the speculation. To the amazement of everybody, Judge Gilbert issued an injunction restraining all parties to all the other suits from further proceedings; in one paragraph ordering the Erie directors to continue in the discharge of their duties — in direct defiance of the injunction of one judge — and in the next paragraph forbidding the directors to desist in the conversion of bonds — in direct defiance of another judge. The Drew interests were now enjoined in every direction. One judge had forbidden them to move, and another judge had ordered them not to stand still.

It was a strategic position which Drew and his agents could not have improved upon, and while matters stood this way the 50,000 shares of Erie stock had been flung on the market. Vanderbilt, who was ignorant of this situation, bought the new stock as eagerly as the old. Then, when the facts came out, the quotations dropped with a thud. Uncle Daniel was victorious; the attempted corner had been a failure; and the Commodore was holding the bag.

Further dramatic events followed. The Erie directors learned that process for contempt had been issued and that their only chance of escape from jail lay in immediate flight. So, stuffing all that was worth while of the Erie Railroad into their pockets, they made off under cover of darkness to Jersey City. One man carried with him in a hackney coach over $6,000,000 in greenbacks. Two of the directors lingered and were arrested; but a majority collected at the Erie station in Jersey City and there, free from interference, went on with the transaction of business. Without disturbance they were able to count their expenses and divide the profits.

Vanderbilt was now loaded up with reams of Erie stock at high costs, and the load was a severe strain on him. He dared not sell for fear of causing a financial collapse. Drew had taken away about seven million dollars of his money and an artificial stringency had been created in Wall Street by this exodus of most of its available cash. But Vanderbilt weathered the storm and, as his generally optimistic attitude inspired confidence, the sky began to clear.

But this stock-market battle did not end the war. New injunctions flew in all directions. Osgood, son-in-law of Vanderbilt, was appointed receiver of the 100,000 shares of illegally issued stock and was immediately enjoined from acting by another judge. Then Peter B. Sweeney, of the Tammany ring, was appointed in his stead without notice to the other side. There was nothing for a receiver to do, as every dollar he was to "receive" was known to be in New Jersey and beyond his reach. Nevertheless he was subsequently allowed a fee of $150,000 by Judge Barnard for his services!

While the legal battle was going on neither Drew nor Vanderbilt was idle. A plot was arranged for bringing the Erie directors over by force, but this failed. In the meanwhile the Erie directors persuaded the New Jersey Legislature to rush through a bill making the Erie Railway a New Jersey corporation. This move, however, was intended merely to meet an emergency. It was the intention of the Erie interests to do their real work with the Legislature at Albany. This was also the intention of the Vanderbilt interests. Consequently, during the subsequent session, the grafters in that body were wooed by both sides. When the Legislature convened, a bill was promptly introduced making legal the recent issue of Erie stock, regulating the power to issue convertible bonds, providing for the guaranty of the bonds of the Boston, Hartford and Erie, and forbidding the consolidation of the Central and the Erie under the control of Cornelius Vanderbilt. But evidently the Commodore's purse was open wider than "Uncle Daniel's," for this bill was defeated by a decisive vote.

Now Jay Gould appeared upon the scene. He left Jersey City with half a million of the Erie's money in his pocket and arrived in Albany immediately after the defeat of this bill. On his arrival he was arrested on a writ issued against him for contempt of court and was held in bail of half a million dollars for his appearance in New York a few days later. He appeared before Judge Barnard in New York and was put in the charge of a sheriff. But the sheriff was served with a writ of *habeas corpus*, and Gould was again brought before the court. Then in some mysterious way the hearing was deferred and Gould returned to Albany, taking the officer as a traveling companion. After reaching his destination Gould became so ill that he could not return to New York, though he managed to go to the Capitol in a driving snow-storm. Here he became rapidly convalescent, as did also many members of the Legislature. Members, indeed, who had been too sick or too feeble to attend the legislative sessions during this cold winter suddenly found their health returning and flocked to Albany on the fastest trains. Gould stayed in Albany until April, and by this time a remarkable change had come over the mentality of a majority of the legislators. On the 13th of April a bill was presented in the Senate which met the approval of the Erie interests and which Judge Barnard afterwards designated as a bill for legalizing counterfeit money. This bill, which was passed after due debate, legalized the issues of Erie bonds and stocks which had been put out by Drew; it provided for the guaranty of the bonds of connecting roads as desired by Drew; and it forbade all possible contracts for consolidation or division of receipts between the Erie and the Vanderbilt roads, a provision also desired by Drew. In fact it was the same bill in different form that had been voted down so decisively a short time before.

But the real tug of war was to get the bill through the lower House. Fabulous stories were told of money which would be expended and the market quotations for votes never soared so high. Then, at the critical moment, Vanderbilt surrendered, made a secret deal with his foe, and withdrew his opposition to the bill. The anger of the disappointed grafters and vote-sellers knew no bounds, and they immediately set to work passing other bills which they felt would annoy or injure Vanderbilt, with the hope that he would still be induced to give them what they regarded as their rightful spoils.

The details of this settlement between Drew and Vanderbilt were not announced until some months afterward. By the terms agreed on Vanderbilt was relieved of 50,000 shares of Erie stock at 70, payable partly in cash and partly in bonds guaranteed by the Erie, and received $1,000,000 in cash for an option given the Erie Railroad to purchase his remaining 50,000 shares at 70 within four months, besides about $430,000 to compensate his friends who had worked so heroically for him. This total sum of nearly $5,000,000 no doubt represented part of the "slush fund" which Drew expected that the company would have to give up to the venal legislators, and it was therefore no hardship to hand it over to Vanderbilt instead.

As a part of the general settlement the Boston

interests were relieved of their $5,000,000 of largely worthless bonds of the Boston, Hartford and Erie Railroad, for which they received $4,000,000 of Erie securities. Thus in all about $9,000,000 in cash or securities was drawn out of the Erie treasury in final settlement of this great stock-market manipulation. And this does not include the pickings of Gould and Fisk and the smaller fry, of which there is no official record. But that these gentlemen did not go empty-handed there is not the shadow of a doubt!

The sensational stock-market deal between the Drew and Vanderbilt interests was but a truce, however, and did not settle the troubles of the Erie. Jay Gould was now becoming a dominating factor and in October of 1868 was chosen president. The various stock-market struggles that ensued from the ascendency of Jay Gould to the receivership of the Erie in 1875 is a long and intricate tale. Suffice it to say that the events were generally similar to those already recounted — stock-market corners, over-issues of bonds and stocks, injunctions, court orders, arrests, legislative bribes. Less than a week after his election Jay Gould frankly announced that the company had just issued

$10,000,000 of convertible bonds and that a third
of these had already been converted into stock.
He further announced that the company now had
$60,000,000 of common stock outstanding, where-
as the public had understood that it was only
$45,000,000.

During the few years that followed, the poor
Erie was systematically looted. Millions were
wasted in New York real-estate speculation, and
the company's money was used in the erection of
the Grand Opera House on Twenty-third Street,
to which the executive offices of the Erie Railroad
were moved. Finally the new ring, comprising as
leading spirits Jay Gould and James Fisk, Jr.,
eliminated Daniel Drew and left him high and
dry without a cent, through a new stock corner.
About this time the road was financially on its last
legs, and Jay Gould was appointed receiver. This
started further litigation which dragged on for
several years until, in 1874, Gould was turned out
by General Daniel E. Sickles in combination with
the English shareholders. The new interests, when
they finally got control, elected an entirely new
management and made H. J. Jewett, a practical
railroad man, president. But the Erie was already
bankrupt, and not much could be done toward

saving the situation. In May, 1875, the road confessed inability to meet its obligations, and Jewett was appointed receiver.

It was three years from the date of the receivership before the Erie property was taken out of the hands of the courts. In April, 1878, a new company, the New York, Lake Erie and Western Railroad, took over the property; Jewett was elected its president, and a new chapter in the history of the property began.

Had the reorganization of the Erie been drastic enough, the road might not so soon have fallen into financial difficulties again, for it owned valuable coal lands in Eastern Pennsylvania and rapidly increased its earnings in this region. Moreover the extension of the system westward should have increased its earning capacity. Up to this time the Erie had no Chicago connection and was at an obvious disadvantage compared with its competitors. It improved this situation in 1881 by acquiring the New York, Pennsylvania and Ohio, and the franchise of the Chicago and Atlantic Railway. Two years later it obtained control of the Cincinnati, Hamilton and Dayton and found itself in a position in which it could compete for through traffic with the Pennsylvania and the New York Central.

But in carrying through these extensive plans, the Erie again became involved in financial difficulties; the sensational Grant and Ward failure in Wall Street in 1884 was a severe blow to the company's credit, as this firm was at that time doing important financing for the Erie. The English security holders stepped to the front again, demanded President Jewett's resignation, and elected John King in his stead.

In 1885 and 1886 a financial readjustment took place, but the company continued to carry the bulk of the heavy load of obligations which had been created during the years of the Drew and Gould managements. It was surely an evidence of the inherent worth of the property that during the half dozen or more years following, the Erie succeeded in struggling along in the face of all its financial and other handicaps and at the same time showed substantial growth in the volume of its business. The company was kept above water until 1893 without again appealing to the courts; but by that time the indebtedness had once more mounted, and in July of that year Erie receivers were appointed for the fourth time in its history.

The name of Pierpont Morgan is closely identified with the story of the railroad during this latest reorganization period. Morgan's firm came to the front in 1894, with the powerful backing of the large English interests, and proposed a plan which involved heavy sacrifices by many of the security holders but which was designed to insure the permanent future of the property. The plan was vigorously opposed, however, by Edward H. Harriman, August Belmont, and other powerful interests, and it was not until August, 1896, that a final compromise was effected and a reorganization was carried through. But at last the Erie was taken out of receivership, and an entirely new company, intelligently designed and having ample working capital for future development, was formed with E. B. Thomas at its head. This new president, like Daniel Willard of the Baltimore and Ohio and many of the modern railroad leaders, was a practical railroad man who had worked up from the ranks and who had no large financial interest or banking connections to divert his attention from the real business of management. Under Thomas, who remained at the head of affairs from 1896 to 1900, the Erie made substantial progress. The system was solidified and its territory was more uniformly and systematically developed. In 1898, the Erie secured control of the New York,

Susquehanna and Western system, gaining thereby
an important branch to Wilkesbarre; and in 1901
it purchased jointly with the Lehigh Valley Railroad the stock of the Pennsylvania Coal Company
of which the Erie later became sole owner. The
real achievement of the Thomas administration was
the development of the property as a heavy carrier of anthracite coal. On the financial side during this period the credit of the House of Morgan,
intelligent administration, and modern methods
did much to improve the reputation of the Erie
and enable it to live down its bad inheritance.

In 1901 Frederick D. Underwood succeeded
Thomas. Like his predecessor, Underwood represented the modern type of railroad president —
a hard-working, eminently practical big business
manager of great executive talent. Underwood's
idea was to make the Erie a great freight-carrying
system by developing its tonnage and its freight
capacity in every way possible. Consequently he
favored opening up the property more extensively in the soft coal fields of Ohio and Indiana, reconstructing roadbeds, laying extra tracks, and
eliminating grades and curves.

The history of the Erie Railroad ever since 1901
has been a record of progress. During these years

he system has been practically rebuilt. It now has a double track from New York to Chicago; it has extensive mileage in the soft coal regions of Ohio and Indiana, and its soft coal tonnage today far overtops its tonnage of anthracite coal; its train load averages far higher than that of the New York Central or of any other Eastern trunk lines except the Pennsylvania; its steep grades throughout New York State have been for the most part eliminated, and many short cuts for freight traffic have been built.

In carrying through these extensive developments in fifteen years the Erie has spent hundreds of millions of dollars. More money indeed has been used legitimately for improvement and development since the reorganizaton of 1896 than during the previous sixty years of its existence. Of course this outlay has meant that the Erie has had to create new mortgages and borrow many millions; but a large part of the expenditure for improvement has come directly from earnings. The Underwood administration has been conservative in paying dividends and the stockholders grumble. But the Erie is at last coming into its own. Instead of being a speculative football and a hopelessly bankrupt road, as it was for nearly forty years, it is now in the forefront of the great trunk lines of the eastern section of the United States. It is no longer, what it was called for many years, the "scarlet woman of Wall Street," but is a respectable member of the American railroad family.

currents up the Mississippi from New Orleans and were threatening the extinction of the aggressive flatboat traffic. Great strides had also been made in the construction of turnpike roads. The famous National Pike from Cumberland to Vandalia, Illinois, had been in large part completed and had done much for the opening up of the Western territory.

Canal building was likewise an extensive development of this period. The idea of connecting the waters of the Chesapeake with those of the Ohio had been broached by George Washington before the Revolution, and he had also prophesied the union of the Hudson and Lake Erie by canal. He believed that a country of such great geographical extent as the United States could not be held together except by close commercial bonds.

The opening of the Erie Canal to New York in 1825 stimulated other cities on the Atlantic seaboard to put themselves into closer commercial touch with the West. This was especially true of the city of Baltimore. A canal connecting Chesapeake Bay and the Ohio River was advocated to protect the trade of Baltimore and the South from the competition of New York and the East which would inevitably result from the construction of the Erie Canal and the Public Works o.

Pennsylvania. But discouragements in plenty frustrated the plan. The cost was believed to be excessive and the engineering difficulties were said to be almost insuperable. George Bernard, a French engineer, was of the opinion that the high elevations and scarcity of water along the route would prevent such a canal from having much practical value. For these reasons Baltimore believed that its position as a center for the rapidly developing Western trade was slowly but surely slipping away.

This was the situation that led to the building of the Baltimore and Ohio Railroad. Two men — Philip E. Thomas and George Brown — were the pioneers in this great undertaking. They spent the year 1826 investigating railway enterprises in England, which were at that time being tested in a comprehensive fashion as commercial ventures. Their investigation completed, they held a meeting on February 12, 1827, including about twenty-five citizens, most of whom were Baltimore merchants or bankers, "to take into consideration the best means of restoring to the city of Baltimore that portion of the western trade which has lately been diverted from it by the introduction of steam navigation and by other causes." The outcome was an application to the Maryland Legislature for a charter for a company to be known as "The Baltimore and Ohio Railroad Company" having the right to build and operate a railroad from the city of Baltimore to the Ohio River. The formal organization took place on April 24, 1827, with Philip E. Thomas as president and George Brown as treasurer. The capital of the proposed company was fixed at five million dollars.

The construction of the railroad began on July 4, 1828. The venerable Charles Carroll of Carrollton, then more than ninety years old and the only surviving signer of the Declaration of Independence of fifty-two years before, said on this occasion, as he laid the first stone: "I consider this among the most important acts of my life, second only to my signing the Declaration of Independence." His vision was indeed prophetic.

It was determined that the first section of road constructed should extend to Ellicott's Mills, twelve miles distant, but, owing to delays in obtaining capital, the actual laying of the rails was not begun until the fall of 1829, and this first section was not opened for traffic until May 22, 1830. At first, experiments were made with sails for propelling the cars, but it was soon found that a more effective source of power was supplied by mules and horses. The *Flying Dutchman*, one of the cars devised to furnish motive power, provided for the horse or mule a treadmill which would revolve the wheels and make the distance of twelve miles in about an hour and a quarter. Steam locomotives at this time were in their infancy and, until the opening of the Liverpool and Manchester Railroad in this same year, they had attained a speed of only six miles an hour. Horses and mules, and even sail cars, made more rapid progress than did the earliest locomotive. In spite of these crude and primitive facilities for transportation, however, the traffic on the new railroad was of large volume from the beginning, and the company could not handle the amount of merchandise offered for transport in the first months.

Construction was now rapidly pushed ahead, and by 1832 the whole line had been opened to Point of Rocks, with a branch to Frederick, Maryland, making seventy-two miles in all. In 1831, steam locomotives were tested, and one of them, the *York*, was found capable of conveying fifteen tons at the rate of fifteen miles an hour on level portions of the road. This achievement was regarded as a great triumph, and in 1832 the directors of the road called attention to "the great increase in velocity" that had been obtained in this way.

From this time forward the expansion of the railroad proceeded with a certainty born of success. A branch was built to Washington and the main line was extended to Harper's Ferry. Beyond this point construction was slow because financial difficulties stood in the way, and it was not until after the panic of 1837 that further aggressive building began. But by 1842 the line was completed to Cumberland, Maryland, and by 1853 to Wheeling. Meanwhile, the branch from Cumberland to Parkersburg, Virginia, was built. The road now comprised a total system of more than five hundred miles and reached two points of importance on the Ohio River, one northward near the Pennsylvania-Ohio state line and one southward in the direction of Cincinnati. The Parkersburg extension was of great importance because it opened a through route to St. Louis, by means of the Cincinnati and Marietta Railroad — which was at this time completed from Cincinnati to Belpre, Ohio, opposite Parkersburg — and the Ohio and Mississippi, which extended more than three hundred miles from St. Louis to Cincinnati.

Times were not the best, however, and, although

LOCOMOTIVE "JOHN BULL," 1831

Photograph from the original engine in the National Museum, Washington. This locomotive was built in Newcastle, England, and brought to America for the Camden and Amboy Railway in August, 1831. It was exhibited at the World's Fair in Chicago in 1893.

much traffic was developed, the immense cost of
the extensions heavily burdened the Baltimore
and Ohio Company, while the panic of 1857
seriously embarrassed its credit. Soon after this
panic and before the company had begun to recover from its effects, John W. Garrett, one of the
large stockholders in the road and son of a Baltimore banker, was elected to its presidency, and a
new chapter in the history of the Baltimore and
Ohio began. Almost immediately following Garrett's election, a remarkable change became apparent. Losses were turned into gains; deficits were
converted into surpluses; and soon Garrett had
gained the reputation of being the most remarkable and efficient railroad manager in the world.
He seemed to be almost an Aladdin of railroad
management for, even when he could not show increases in amount of business done, he reported
greater profits by showing lower expenses. In
those days the railroads did not furnish detailed
reports of business to the stockholders or to the
public. At the annual meetings it was customary
for a president or the directors simply to an-
nounce, either orally or in a brief printed statement, the amount of gross business and profits for
the year. No such thing as a balance sheet or

Pittsburgh, Sandusky, and Chicago, and further strengthened its connections with Cincinnati and St. Louis. It acquired steamboats, grain elevators, and docks; it constructed hotels as mountain summer resorts; it built dry docks in Baltimore; and finally it proceeded to organize and operate an express company, a telegraph company, and a sleeping-car company. To carry out these ambitious plans the capital stock and debt were of course increased again and again, and in the course of these operations a large part of the new securities issued was sold to English investors. Notwithstanding these great increases in liabilities, the company continued to report large surpluses and to pay large dividends, — generally ten per cent annually. In fact, this liberal rate was, with brief exceptions, paid right through the Civil War period, in spite of the fact that large parts of the line were frequently destroyed and traffic was often at a standstill. With such prosperity under such conditions Garrett's reputation as a railroad manager naturally suffered no eclipse.

In the course of the Civil War, as already noted, through traffic routes from New York to Chicago had been established, and in the succeeding years the consolidations of the great competing systems into trunk lines had taken place. The struggle of the Baltimore and Ohio for its share of Western business led to fierce rivalry with the Pennsylvania. This competition became so severe and intense that, in 1874, the Pennsylvania road refused to carry the Baltimore and Ohio cars over its line to New York on any terms whatever. Since this was the only way in which the Baltimore and Ohio could reach New York, the situation was a serious one. Garrett retaliated by making destructive reductions in passenger rates from Washington and Baltimore to Western points. The cuts were soon made on other roads and affected both freight and passengers. All the lines became involved. Passenger fares from Chicago to Baltimore and Washington were reduced from nineteen dollars to nine dollars, and those to New York and Boston from twenty-two to fifteen dollars. Still the fight continued, and before the end of 1875 it was possible to travel from Chicago to New York first class for twelve dollars and to ship grain to New York for as low a rate as twelve cents.

Despite the fact that competition had cut earnings almost to the point of extinction, the Baltimore and Ohio continued to report surprisingly good profits. The company borrowed additional

unds from time to time but continued to pay the
liberal ten per cent dividend until 1877, when it
somewhat reduced the rate. These dividend payments indicated, however, a prosperity that was
only apparent, and they did not greatly deceive the
bankers, for the credit of the Baltimore and Ohio
weakened from day to day. The fact is that the
reports of operations inspired little public confidence; to the farseeing, there were danger signals
ahead. Nevertheless the ten per cent dividends
were resumed in 1879 and continued at this rate
without interruption until 1886.

On the death of John W. Garrett in 1884, his son
Robert, who succeeded him as president, continued
the same policy of competition and aggression.
With the object of gaining an entrance into Philadelphia and through that gateway of reaching New
York, he started work on a branch from Baltimore
to Philadelphia to meet, at the northern boundary
of Maryland, the Baltimore and Philadelphia Railroad — a line which independent interests were
then building through Delaware with the intention
of obtaining an entrance into Philadelphia. The
Pennsylvania interests strongly opposed Garrett's
new project and many years before had gone so
far, in their determination to block the Baltimore

and Ohio from acquiring control of the Philadelphia, Wilmington and Baltimore Railroad, as to purchase that road themselves. Despite this opposition the Baltimore and Ohio went forward with their plans and secured an entry into Philadelphia by acquiring control of the Schuylkill East Side Railway, which was a short terminal road of great strategic value. North of Philadelphia the company arranged a traffic contract with the Philadelphia and Reading, whose lines extended to Bound Brook, New Jersey, and also with the Central Railroad of New Jersey beyond Bound Brook to Jersey City. Afterward, by purchasing the Staten Island Rapid Transit Company the Baltimore and Ohio acquired extensive terminals at tidewater on Staten Island and constructed a connection in New Jersey with the New Jersey Central. Thus, after many years of struggle and at heavy cost, the Baltimore and Ohio finally secured an entry into the New York district independently of the Pennsylvania Railroad.

Both freight and passenger charges, however were still maintained at an unprofitable rate, and after the death of John W. Garrett, the credit of the Baltimore and Ohio continued to decline. Dividends were gradually reduced and by 1888 were omitted entirely. As is usually the case, the cessation of dividends awakened the sleeping stockholders. They began an investigation to ascertain the whereabouts of that remarkable surplus which had been reported from year to year and which, according to official report, had shown a constant growth.

This investigation disclosed a startling state of affairs. Instead of a surplus, the company had been piling up deficits year after year, had been borrowing money right and left on onerous terms, had been charging up millions of dollars of expenses to capital accounts — and as a matter of fact, instead of making money, it had for the most part been losing it. Now the company urgently needed cash, and the only way it could obtain that essential commodity was by selling its express, telegraph, and sleeping-car business.

During the entire administration of John W. Garrett, extending over more than two decades, current expenditures of enormous amounts which should have been deducted from the income had been credited to the surplus; many millions which would never be returned had been advanced to subsidiary lines, or had been spent, and therefore should have been put down in the books as losses. When these facts became public, the capital stock of the Baltimore and Ohio, which for generations had been looked upon as one of the most secure of railroad investments, dropped to almost nothing, and the most strenuous financial efforts were required to keep the company out of bankruptcy.

These disclosures, towards the end of 1887, ended the first period of active Garrett management in the Baltimore and Ohio. The directors then turned to New York bankers for the cash that was needed to put the affairs of the company on a sound basis. Samuel Spencer, who afterward became a partner in the banking house of J. P. Morgan and Company, was elected president and active manager. He introduced radical reforms, entirely revolutionized the organization, and adopted modern methods. He wrote off the books a large amount of the much vaunted "surplus" and he took important steps toward the general improvement of the property.

Had the new interests been allowed to continue their efforts unmolested, the history of the Baltimore and Ohio in the next decade might have been very different. But the original controlling interests, the Garrett family, still held the balance of power. As the bad bookkeeping and other irregularities of the past naturally reflected on the

Garretts, it was their interest to suppress further investigation as far as possible; and their antagonistic attitude toward the policy adopted by the new Spencer management was seen in the annual election of directors in November, 1888. Only five of the members of the board were reëlected, President Spencer was ousted, and Charles J. Mayer was elected in his place.

This second change in management sidetracked the plans for radical reform, and little improvement resulted either in earning power or in financial condition. The company had fallen upon evil days. The net profits did not increase, and eight years after 1888 they were smaller than in that year, while the debt and interest charges constantly grew. Despite these ominous facts, dividends were paid regularly on the preferred stock and in 1891 they were resumed on the common stock. In the latter year a twenty per cent dividend was declared "to compensate shareholders for expenditures in betterments and improvements in the physical condition of the property," while at the same time the directors decided to raise five million dollars of new capital for expenditures which would be necessary to handle the increased traffic created by the World's Fair at Chicago.

The traffic problem continued to be a thorn in the flesh and until 1893 freight rates were constantly being cut. The opening of the Baltimore and Ohio connection to New York had brought keener competition from the Pennsylvania Railroad and had made deep inroads into the Baltimore and Ohio revenues. Such conditions made even the Garrett interests feel that something should be done, and in 1890 a "community of interest" scheme was proposed. To control the stock of the Baltimore and Ohio Railroad, Edward R. Bacon in New York, acting harmoniously with the Garrett family, formed a syndicate of capitalists representing the Richmond Terminal system, the Philadelphia and Reading Railroad, the Northern Pacific Railroad, and other properties. The ultimate plan, which proved too visionary, was to consolidate under one control a vast network of lines extending all over the continent.

The syndicate had made little progress toward rehabilitation when the panic of 1893 occurred. In this year and the next the earnings of the Baltimore and Ohio fell off rapidly and the dividend was reduced. Nevertheless, as late as January, 1895, the directors insisted that financially the company was in better condition than for several years and that on the whole it was in a stronger position than at any time since 1880. But in this same year it became necessary to stop all dividend payments; the company began to have difficulties in securing ready money; and before the close of the year the situation seemed hopeless. Early in 1896 Mayer tendered his resignation, and John K. Cowan succeeded him. The new president did his utmost to obtain money to meet the current needs, but he was unsuccessful. A receivership and reorganization seemed absolutely necessary, and in February, 1896, the receivership was announced.

With the property now in the hands of the courts, the opportunity at last came to make real the reforms which had been proposed and begun nearly a decade earlier under the wise but quickly terminated administration of Samuel Spencer. A thorough housecleaning was now carried through without interference or interruption. A reorganization committee was formed, with whom were deposited the Garrett shares as well as those of the Morgan and New York and Philadelphia interests. A full investigation of past management disclosed that the records for the interim extending from the brief Morgan control under Spencer to the receivership contained the same kind of irregularities and errors of policy that had prevailed under the earlier Garrett management. Statements of profits had been swelled by arbitrary entries in the books and nearly six million dollars which had not been earned had been paid out in dividends. Furthermore the company had endorsed the notes of certain subsidiary roads to the extent of over five million dollars, and had made no record whatever of this action for the stockholders.

As in the case of numerous other railroads, the financial breakdown of the Baltimore and Ohio Railroad was primarily due to a bad or reckless financial policy, for there was nothing inherently insecure in the railroad property itself. During all the years of the Garrett régime, the company had shared in the general growth and expansion of industry, wealth, and population within its territory. It had been progressive in matters of expansion and had built up its system to meet the needs of modern times. Its trackage and equipment compared favorably with similar systems, and most of its extensions and branches had been wisely planned and had proved profitable. The operating management of the railroad was generally good and it usually secured its proportion of what business was to be obtained. But the steady increase in its debts over a number of years, its extravagance in dividend payments, and its painful efforts to keep down its operating expenses had so weakened the property that, when the hard times of 1893 to 1896 arrived, it was in no position to weather the storm. The only wonder is that the management succeeded in keeping the system intact and apparently solvent so long as it did.

The receivership at once adopted a vigorous policy of improvement. The rolling stock had run down until it could not handle even ordinary business. While the company had been depleting its credit and paying out all its cash in dividends, the equipment had been going into the scrap heap. For two years the receivers made large expenditures on equipment and roadbed, borrowing money for this purpose; the result was that when, in 1898, the courts surrendered the property, it was in splendid condition to take advantage of the tide of commercial and industrial prosperity which was just then beginning to flow throughout the United States.

While the reorganization of the Baltimore and Ohio was not so drastic as that of many other systems which went through the courts during this period, it was thorough enough to meet the

8

much attention to the improvement and development of terminals; and they spent many millions in acquiring and expanding the terminal properties of the company at Chicago, St. Louis, Philadelphia, and Baltimore.

The financial history of the Baltimore and Ohio since the close of the nineteenth century is interesting chiefly in connection with changes in the control of the property. After the reorganization a group of prominent financiers, including Marshall Field, Philip D. Armour, Norman B. Ream, and James J. Hill jointly purchased a large interest in the stock. But this purchase, while perhaps representing a dominating interest, did not involve actual control. Soon afterward, interests identified with the Pennsylvania Railroad began to appear in the Baltimore and Ohio, and before long the Pennsylvania had a strong representation on the board. As a consequence, the Baltimore and Ohio almost lost its individuality and for a time was popularly regarded practically as a subsidiary of its old rival line.

The purpose of the Pennsylvania in obtaining his ascendency over the Baltimore and Ohio was to regulate the soft coal traffic. Already it had acquired dominating interests in the Chesapeake and

Ohio, the Norfolk and Western, and other soft coal properties. These purchases were merely manifestations of that "community of interest" policy which at this time led several large systems to acquire interests in competing lines. Several of the railroad leaders of that time, notably James J. Hill and Edward H. Harriman, believed that if these great systems actually owned large blocks of stock in each other's properties, this common association would *ipso facto* end the competition that, if continued, would ultimately ruin them all. The Supreme Court had decided that the "pooling" arrangements which had so long prevailed among great competing roads violated the Sherman Anti-Trust Act; and the American public, which now was cultivating a new interest in railroad problems, believed that the "community of interest" plan was merely a scheme to defeat the Interstate Commerce Act and the Sherman Act and to maintain secretly all the old railroad abuses. These inter-railroad purchases therefore became so unpopular that the Pennsylvania sold its Baltimore and Ohio stock. At this time Edward H. Harriman of the Union Pacific, who had at his disposal vast funds of the latter property which he had obtained by the settlement of the Great Northern and Northern Pacific deal, decided to acquire control of a system of roads in the East in order to establish a complete transcontinental line in the interest of the Union Pacific. It was the theory that such a purchase by the Union Pacific would not defy the law or outrage the popular conscience because the Union Pacific, unlike the Pennsylvania, did not compete with the Baltimore and Ohio, but was only a western extension of that system. Harriman in August, 1906, therefore purchased nearly all the Pennsylvania holdings in the old Garrett property and thus obtained virtual control.

At this same time the Baltimore and Ohio had been developing a "community of interest" plan on its own account. In the year 1903, it acquired a substantial stock interest in the newly reorganized Reading Company, which controlled the Philadelphia and Reading Railroad and the Philadelphia and Reading Coal and Iron Company. It did not obtain a majority interest but, with the Lake Shore and Michigan Southern Railroad of the New York Central system, it now controlled the Reading system. The Reading Company meanwhile had secured control of the Central Railroad of New Jersey, over the lines of which the Baltimore and Ohio reached New York City.

In the following years the Baltimore and Ohio property was still further rounded out by purchasing the Cincinnati, Hamilton and Dayton, a small system of doubtful value radiating through the State of Ohio and, by additional extensions, into the soft coal fields of West Virginia. New energy was put into the expansion and improvement of the southwestern lines to St. Louis, while the eastern terminal properties were still further improved.

The practical control of the Baltimore and Ohio remained in the hands of the Union Pacific interests until 1913. In that year, however, the Union Pacific liquidated its holdings by distributing them to its own individual stockholders in the shape of a special dividend. The Baltimore and Ohio thus became once more an independent property.

The story of the Baltimore and Ohio for the past decade has been mainly a record of a growing, well-managed, and efficient business. It is closely identified with the personality of its notable and efficient president, Daniel Willard, a conspicuous example of the modern type of railroad manager. In the earlier days of railroading, and especially in the long period which came to an end with the death of Harriman, the typical railroad president was usually a man of great wealth who had secured his position by owning a large financial interest in the property. The country was full of "Wall Street Railroad Generals." But in recent years the efficient railroad head has come more and more to be the practical railroad man who has risen from the ranks, who has no important personal financial interest in the property but who is paid an adequate salary to operate a system in a purely businesslike way. Notable examples of this modern type of railroad president are, besides Daniel Willard, Edward P. Ripley of the Atchison, Topeka and Santa Fé, Benjamin F. Bush of the Missouri Pacific, and Fairfax Harrison of the Southern.

The efficient management of today is abundantly shown in the recent record of the Baltimore and Ohio. President Willard has been unmolested by financial interests and has been continuously backed up in his policies by the owners of the road. As a result the Baltimore and Ohio of the present decade has reached an enviable position as one of the great Eastern trunk lines, comparing well with other progressive properties like the Pennsylvania, the New York Central, the Southern, the Illinois Central, and the Louisville and Nashville. Mil-

CHAPTER VI

IN 1862, when the charter was granted by the United States Government for the construction of a railroad from Omaha to the Pacific coast, the only States west of the Mississippi Valley in which any railroad construction of importance existed were Iowa and Missouri. During the three decades which had passed since the first railroad construction, the earlier methods of transportation by boat, canal, and stage coach gave place in the Eastern half of the United States to more modern methods of transportation. As a result of these new conditions, the States, cities, and towns were welded together, and population and prosperity increased rapidly in those inland sections which had formerly languished because they had no means of easy and rapid communication.

The construction of extensive railways, however, and particularly the consolidation of small, experimental lines into large systems, dates from the days of the discovery of gold in California. The nation did not begin to realize the extraordinary possibilities of the vast Western territory until its attention was thus suddenly and definitely concentrated on the Pacific by the annual addition of over fifty million dollars to the circulating medium. The wealth drawn so copiously from this Western part of our continent had a stimulating effect on the commerce, manufactures, and trade of the entire Eastern section. People began to understand that with the acquisition of California the nation had obtained practically half a continent, of which the future possibilities were almost unlimited, so far as the development of natural resources and the general production of wealth were concerned.

The public conviction that a railroad linking the West and the East was an absolute necessity became so pronounced after the gold discoveries of '49 that Congress passed an act in 1853 providing for a survey of several lines from the Mississippi to the Pacific. Though the published reports of these surveys threw a flood of light on the interior of the continent, they led to no definite result at the time because the rivalry of sections and groups of interests for the selection of this or that route held up all progress.

The Act of 1862, which created the Union Pacific Railroad Company, together with the amending Act of 1864, authorized the construction of a main line from an initial point "on the one hundredth meridian of longitude," in the Territory of Nebraska to the eastern boundary of California, with branch lines to be constructed by other companies and to radiate from this initial point to Sioux City, to Omaha, to St. Joseph, to Leavenworth, and to Kansas City.[1] Provision was made for a subsidy of $16,000 a mile for the level country east of the Rocky Mountains; $48,000 a mile for the lines through mountain ranges; and $32,000 a mile for the section between the ranges. The original plan to secure the government subsidies by a first mortgage on the lines was amended so as to allow private capital to take the first mortgage, the Government taking a second lien for its advances. In addition to these subsidies the several companies were to receive land grants of 12,800 acres to the mile in alternate sections contiguous to their lines. Upon the same terms the Central

---

[1] These ambitious designs were never fully realized. The main line ran eventually west from Omaha, meeting the Sioux City branch at

Pacific, a company incorporated under the laws of California, was authorized to construct a line from the Pacific coast, at or near San Francisco, to meet the Union Pacific Railroad.

The public was quick to realize the significance of this huge enterprise, for the papers of the day were full of such comments as the following:

It is useless to enlarge upon the value and importance of this great work. It concerns, not the United States alone, but all mankind. Its line is coincident with the natural and convenient route of commerce for the world. . . . Over it the trip will be made from London to Hong Kong in forty days, over a route possessing every comfort and attraction, which takes a continent in its course, and which, from the variety and magnitude of its sources, from the race which now dominates it, and from the extent of their numbers, wealth and productions, must soon give law to the commercial world.

Notwithstanding these and similarly optimistic sentiments, the meager financial support given to the enterprise by the public at large had been very discouraging. Although the construction had been liberally subsidized by the Government, gross extravagance had promptly crept in; juggling of accounts for the purpose of securing profits on the

---

Frémont. The only other branch which was constructed to connect with the Union Pacific was that from Kansas City and it ran first to Denver.

government advances was freely indulged in, and after only a small section of the line had been completed it was announced that more capital must be forthcoming or the work would cease. Out of this situation grew the plan for subletting the work to a construction company known as the Pennsylvania Fiscal Agency — a name which was afterwards changed to that of the Crédit Mobilier of America. The story of the Crédit Mobilier, with its irregularities involving conspicuous politicians, is one of the most disgraceful in American history. The detailed history of these operations need not be considered here; it is sufficient to say that finally, in spite of political scandals, the Union Pacific lines were brought to completion. Within two years after the letting of the contracts to this new company, in 1866, over five hundred miles of road were completed and in operation. An advertisement published late in 1868 announced that "five hundred and forty miles of the Union Pacific Railroad, running west from Omaha across the continent, are now completed, the track being laid and trains running within ten miles of the Rocky Mountains. . . . The prospect that the whole grand line to the Pacific will be completed by 1870 was never better."

As a matter of fact, the line through to the coast was finished earlier than had been predicted. One fact which increased the rapidity of construction was the growing financial difficulty of the company. It was absolutely imperative that the through line be completed in order that the resulting business might make the operation of trains pay. But aside from this, another influence was at work to encourage rapid construction. The Act of 1862 provided that the Central Pacific might also build across Nevada to meet the Union Pacific, on condition that it completed its own allotted section first. As the Central Pacific also was receiving a heavy government subsidy per mile, and as there was great profit in construction undertaken with this government subsidy, there was naturally a strong incentive for both companies to build all the mileage possible and as rapidly as possible.

The Central Pacific enterprise was backed by a group of men who were awake to the possibilities of the situation and who had made large fortunes in the gold-mining boom of previous years, such as Leland Stanford, Collis P. Huntington, Mark Hopkins, and the Crockers. The rivalry between them and the Union Pacific interests woke the whole continent and formed a chapter in

American railroad history as startling and romantic as anything in the stories of the Vanderbilts and Goulds with their financial gymnastics.

As the contest proceeded, public interest increased and the entire country watched to see which company would win the big government subsidies through the mountains. Through the winter of 1868 the work continued on the Union Pacific with unabated energy, and freezing weather caught the builders at the base of the Wasatch Mountains; but blizzards could not stop them. The workmen laid tracks across the Wasatch on a bed of snow and ice, and one of the track-laying trains slid bodily, track and all, off the ice into a stream. The two companies had over twenty thousand men at work that winter. Suddenly the Central Pacific surprised the Eastern builders by filing a map and plans for building as far as Echo, some distance east of Ogden. The Union Pacific forces, however, were equal to the occasion. At first, one mile a day had been considered rapid construction, but now, even with the limited daylight of the winter months, they were laying over two miles a day, and they finally crowned their efforts by laying in one day between sunrise and sunset nearly eight miles of track.

In the meantime the Central Pacific also had stopped at nothing. The company had a dozen tunnels to build but did not wait to finish them. Supplies were hauled over the Sierras, and the work was pushed ahead regardless of expense. On May 10, 1869, the junction was formed, the opposing track layers meeting at Promontory Point, five miles west of Ogden, Utah. Spikes of gold and silver were driven into the joining tracks, and the through line from the Missouri River to the Pacific Ocean had been completed; the first engine from the Pacific coast faced the first engine from the Atlantic. The whole country, from President Grant in the White House to the newsboy who sold extras, celebrated this achievement. Chicago held a parade several miles long; in New York City the chimes of Trinity were rung; and in Philadelphia the old Liberty Bell in Independence Hall was tolled again.

The cost of the Union Pacific Railroad from Omaha to its junction with the Central Pacific formed a subject of controversy for a generation. The saving of six months of the allotted time for completing the road no doubt increased its cost to the builders, for at times they borrowed money in the East at rates as high as 18 and 19 per cent.

Besides, in pushing the line far beyond the bounds of civilization without waiting for the slower pace of the settler and the security which his protection afforded, it often became necessary for half the total number of workmen to stand guard and thus reduce the working capacity of the construction force. Even so, hundreds were killed by the Indians. Governmental restrictions of various kinds also increased the cost of the road. For example, the stipulation that only American iron should be used increased the cost by at least ten dollars for every ton of rail laid. The requirement that a cut should be made through each rise in the Laramie plains, thus giving the track a dead level instead of conforming to the natural roll of the country, ultimately resulted in a waste of from five to ten million dollars. Extraordinary costs such as these, combined with the extravagant methods of construction and financing, brought the total cost of the property up to what was in those days a fabulous sum of money. The records indicate that the profits which accrued through the Crédit Mobilier and in other ways in the construction up to the time of the opening in 1869 exceeded fifty millions of dollars.

While the Union Pacific was being built, from

9

1862 to 1869, other railroads were not idle, and many were rapidly reaching out into the Central West. Not only had the Chicago and North Western reached Omaha and made connection with the Union Pacific, but the Kansas Pacific had penetrated as far west as Denver and had joined the Union Pacific at Cheyenne.

The close relationship between railroad expansion and the general development and prosperity of the country is nowhere brought more distinctly into relief than in connection with the construction of the Pacific railroads. With the opening of a transcontinental line the vast El Dorado of the West was laid practically at the doorstep of Eastern capital. Not only did American pioneers turn definitely toward the West, but foreign emigrants bent their steps in vast numbers in that direction, and capital in steadily increasing amounts made its way there. Towns sprang up everywhere and soon developed into busy centers of trade and commerce. Caravan trains, which a few years before had followed a single westward line, now started from points along the railroad artery and penetrated far to the north and south. The settlers knew that the time was not far distant when all the vast territory west of the

Missouri, from the Canadian border to the Rio Grande, would be reached by the rapid spread of the railroad. In the sixties and seventies there sprang up and rapidly developed in size and importance such centers as Kansas City, Sioux City, Denver, Salt Lake City, Cheyenne, Atchison, Topeka, Helena, Portland, Seattle, Duluth, St. Paul, Minneapolis, and scores of smaller places. The entire Pacific slope was soon dotted with towns and cities, and even the great arid plains of the West — as well as the "Great American Desert" covering Utah, Arizona, New Mexico, and parts of Nevada — began to take on signs of life which had not been dreamed of a decade before.

But the development of this great section of the country during the next few years was even more notable. By 1880 four different lines of railroad were running through to the Pacific States, and a fifth, the Denver and Rio Grande, had penetrated through the mountains of Colorado and across Utah to the Great Salt Lake. These were the years when the modern industrial era was really beginning. Man's viewpoint was changing, and instead of remaining content with the material achievements of the Atlantic and Central sections of the continent, he began to realize that the vast Western regions and the thousand miles of Pacific coast line were destined to be America's inexhaustible patrimony for the years to come.

In 1880 the Union Pacific began its expansion to the eastward and acquired control of the Kansas Pacific, which had come upon evil days, and of the Denver Pacific, a most important connecting link. In January, 1880, these two companies were absorbed by the Union Pacific, which thus obtained a continuous line from St. Louis westward. In the meantime the Central Pacific, operating from Ogden west to the coast, had added many branches, while a new company — known as the Southern Pacific Railroad of California — had for some years been constructing a system of lines throughout that State south of the Central Pacific and by 1877 had penetrated to Yuma, Arizona, 727 miles southeast of San Francisco. It had also built lines into Arizona and New Mexico and soon joined the Santa Fé route, which had for some time been working westward.

During 1881 the Southern Pacific continued its eastern extensions along the Rio Grande to El Paso, Texas, where it formed a connection with a new road under construction from New Orleans. A junction was also made at El Paso with the

COMPLETION OF THE TRANSCONTINENTAL
RAILROAD

The locomotives *Jupiter*, of the Central Pacific, and *119*, of the Union Pacific, meeting on May 10, 1869, at Promontory Point, Utah, when the last spike was driven. Drawing from a wood engraving published in 1870, reproduced in *The History of Travel in America*, by Seymour Dunbar.

Mexican Central, which was under construction to the City of Mexico. The Southern Pacific Railroad was closely allied with the Central Pacific interests headed by Collis P. Huntington, and in 1884 the great Southern Pacific Company was formed, which acquired stock control of the entire aggregation of railroads in the South and Southwest. At the same time the Central Pacific came under direct control of the Southern Pacific through a long lease.

During these eventful years, while the Southern Pacific properties were penetrating eastward through the broad stretches of country to the south of the Union Pacific lines, equally interesting events were occurring in the north. In 1879 a consolidation was formed of the Oregon Steamship and Navigation Company with several short railway lines in Oregon and Washington, under the name of the Oregon Railway and Navigation Company. These railroad lines extended east from Portland to the Oregon state line, and north to Spokane, and they finally made connection with the new Northern Pacific. At the same time, another road, known as the Oregon Short Line Railroad, was built from Granger, Wyoming, on the line of the Union Pacific to a junction with

the Oregon Railway and Navigation Company at
Huntington, Oregon, on the Snake River. The
Oregon Short Line came under the control of the
Union Pacific and was opened for traffic in 1881.
Later a close alliance was made with Henry Villard, the controlling spirit in the Oregon Railway
and Navigation Company. Ultimately the entire
system of Oregon lines passed under Union Pacific
control, to be lost in the receivership of 1893, but
later recovered under the Harriman régime.

When, after ten more years of expansion, the
great Union Pacific property went into the hands
of receivers in 1893, it had grown to a system of
more than 8000 miles. It completely controlled
the Oregon railway and steamship lines, the lines
to St. Louis, and also an important extension
known as the Union Pacific, Denver and Gulf
Railroad, running from a point in Wyoming across
Colorado to Fort Worth, Texas. The financial
failure of the system was due to a variety of causes.
Its management had been extravagant and in-
efficient, and construction and expansion had been
too rapid. The policy of building expensive branch
lines where they were not needed and of obligat-
ing the parent company to finance them had been
a grievous mistake and had contributed largely to

he downfall of the company. Further than this,
he credit of the Union Pacific was steadily growing
weaker because the time was drawing near when
s heavy debt to the United States Government
would fall due. In all its history of more than
twenty years the company had never paid any
interest on the government debt nor had it maintained a sinking fund to meet the principal when
due. Consequently, the accruing interest had
mounted year by year and, should the Government enforce payment at maturity in 1897–99,
the company would be doomed to bankruptcy.
This government debt, including accrued interest,
amounted to the sum of $54,000,000.

Attention should not, however, be diverted from
the fact that during all these years a vast expansion of competitive lines had been going on far
southward of the Union Pacific. Under the guid-
ing genius of Collis P. Huntington, the Southern
Pacific Company in 1884 had consolidated and
solidified a gigantic system of railways extending
from New Orleans to the Pacific and throughout
the entire State of California to Portland, Oregon,
with branch lines radiating through Texas and
making close connection with roads entering St.
Louis. In addition to these railroads, Huntington

acquired control of a steamship line operating fro
New York to New Orleans and Galveston, a
subsequently of the Pacific Mail Steamship Cor
pany, operating along the coast from Oregon sou
to the Isthmus of Panama and across the Paci
Ocean. The ever-growing effects of this powe
ful and well-managed competitor — combined wi
the large development of the Santa Fé syste
during these years, the competition of the co
pleted Northern Pacific, and the possibilities
the new Great Northern Railway or Hill line, no
completing its main artery to the Pacific — we
far-reaching enough in themselves to bring t
Union Pacific upon evil days. Consequently f
were surprised when, under the great pressure
the panic of 1893, the property was forced
confess insolvency. The Union Pacific had simp
repeated the story of most American railroads;
had been constructed in advance of populati
and had to pay the penalty. Yet it had more th
justified the hopes of the daring spirits who p
jected it. It may have made individuals ba
rupt, but it magnificently fulfilled the part wh
it was expected to play. It had opened up n
lions of acres to cultivation, given homesteads
millions of people, many of whom were immigra

rom Europe, developed mineral lands of incalcu-
able value, created several new great States, and
nade the American nation a unified whole.   Its
ubsequent history belongs to another chapter of
his story — a history that is richer than the first
n the matter of financial success but that can
ıever surpass the early pioneering years in real
ınd permanent achievement.

CHAPTER VII

PENETRATING THE PACIFIC NORTHWEST

It is only when one reads such a book as Franci
Parkman's *Oregon Trail* that one fully realizes th
vast transformation which has taken place withi
little more than half a century in the great North
western territory beyond the Mississippi and th
Missouri. In that fascinating history we read o
the romantic and thrilling experiences of Parkma
and his companions in their summer journey acros
the plains of Nebraska and through the mountai
ranges of Wyoming, Montana, and Oregon. W
read of their hairbreadth escapes from the Indians
their chase of the buffalo and other wild animal
of the far Western country; of the wearisome week
that they spent in crossing the deserts where ab
solute loneliness reigned; and finally of their arri
val, after months of hardship, in the vast Orego
country, which with its great natural resource
splendid climate, and large extent has come to b

known in these modern days as the Empire of the Northwest.

It was to penetrate and bring this great virgin region within reach of the East that the Northern Pacific Railroad Company was chartered by Congress in 1864, just prior to the closing of the Civil War. During this same period the Union Pacific route was being surveyed, and the first ground was broken in December, 1863, for the line which was later to connect Omaha with San Francisco.

Like the Union Pacific charter, that of the Northern Pacific also contained an extensive land grant. From the modern viewpoint, such land grants look colossal, but in those days the general opening up and development of the Western country had progressed to so slight an extent that the significance of giving away millions of acres of the public lands to encourage a precarious railroad enterprise was then no more than the passing over to capitalists today of exclusive rights in extensive tracts of territory in Brazil and the other South American Republics. Even these great opportunities to acquire almost an empire of fertile lands or rich forests were not as a rule looked upon as attractive enough to tempt capital into the wilderness. The old saying that capital is the most

timid thing in the world and does not like pioneering is strongly emphasized by such instances
as this, and no doubt in 1864 the enormous grants
of free land made by Congress did not appear
especially attractive to the man who had money
to invest.

Whatever the public attitude may have been
the Act of Congress of July 2, 1864, creating the
Northern Pacific Railroad, gave that Company
the right to construct a line from some point on
Lake Superior, either in Minnesota or in Wisconsin, westward and north of latitude 45°, to or near
Portland, Oregon. The land grant consisted of
forty alternate sections of public land for each mile
within the Territories penetrated and twenty alternate sections within the States through which
the railroad might pass.

The hazardous character of this undertaking
will be realized when it is remembered that at this
time no railroad had yet penetrated the Rocky
Mountains; that the entire railroad system of the
United States was less than 40,000 miles; and
that west of the Mississippi there was no mileage
worth mentioning. It was still less than a generation since Parkman and his companions had made
their four months' journey from St. Louis to the

mouth of the Columbia River, and between the
fringe of civilization along the Pacific slope and
the region about Chicago and St. Louis lay almost
a third of the continent uninhabited, undeveloped,
and unknown. The scheme languished for several
years until finally, in 1869, the firm of Jay Cooke
and Company of Philadelphia undertook to raise
the necessary capital.

The story of the Northern Pacific for the next
few years was closely bound up with that of Jay
Cooke, who was one of the most conspicuous char-
acters of his time in the financial world. He was
a man of commanding personality, great energy,
unusual resourcefulness, and with a large personal
following. He had built his reputation through his
great success in financing United States government loans during the Civil War. He now undertook to raise more than one hundred million
dollars to carry through the Northern Pacific enterprise. He achieved remarkable success for a
time and within three years had built over five
hundred miles of the main line to the Pacific coast.
But the outbreak of the Franco-Prussian War and
the consequent financial stringency abroad, the
difficulty of marketing bonds on an uncompleted
enterprise, combined with the poor showing made

by those sections of the line completed and in ope
ation, brought matters to a crisis, and in Septen
ber, 1873, Jay Cooke and Company were oblige
to close their doors. The affairs of the railroa
were so closely involved with those of the banl
ing firm that, although strenuous efforts wei
adopted to save the railroad, its revenues wei
inadequate. As a result, in April, 1874, Gener.
Lewis Cass was appointed receiver.

The uncompleted property was operated fc
some years thereafter under the protection of tl
courts and no plan of reorganization was devise
until 1879. During the receivership only a mode
ate amount of additional mileage was constructe
and it was not until many years had passed th;
the system penetrated the mountains and reache
the Pacific coast. But when the new compan
took possession in 1879, aggressive building wi
resumed, and for a time it looked as though tl
project would be promptly finished. Howeve
in 1882, the company still had about one thoi
sand miles to construct in order to complete i
main artery. At this time financial difficulti
appeared, and the days of stress were tided ov
only by the help of a syndicate and the Oregon an
Transcontinental Company.

With the formation of the Oregon and Transcontinental Company begins the régime of Henry Villard, the dominating factor in Northern Pacific affairs for many years afterward. Some years before, Villard, who had long been interested in Western railroad enterprises and who had become prominent through his activities in connection with the Kansas and Pacific Railway, had succeeded in forming the Oregon Railway and Navigation Company as a combination of steamboat lines operating on the Willamette and Columbia rivers in Oregon, with an ocean line connecting Portland and San Francisco. A connecting railroad line, which had been built to Walla Walla in southeastern Washington, penetrated a portion of the territory through which the Northern Pacific was projected. In 1880 a contract was arranged between the two companies whereby the Oregon Railway and Navigation Company, in order to share in the traffic, undertook to construct a line eastward to meet the Northern Pacific line at the mouth of the Snake River. This arrangement would allow the Northern Pacific to run its trains into Portland and would obviate the necessity of constructing its own road into that city.

In spite of this arrangement, Villard feared that

the Northern Pacific Company might decide, after all, to build its own line to Portland as soon as it was able to finance the project. It was for the purpose of preventing this move that he formed the Oregon and Transcontinental Company, a holding corporation which promptly acquired, in the open market and by private purchases, a dominating interest in the Northern Pacific Railroad. At the same time Villard placed the control of the Oregon Railroad and Navigation Company in the hands of the new Transcontinental.

Villard thus came to control the entire Northern Pacific system and, backed by the Deutsche Bank of Berlin and other German and Dutch interests, at once began an aggressive policy of expansion and development. The business of the system developed rapidly. The main line through to the Pacific coast was now in operation, and the entire system amounted to about 2300 miles of road. But Villard followed a financial policy which was not sound and paid dividends without justification. In a short time the company consequently found itself financially embarrassed.

As a result of financial losses in 1884, Villard was obliged to retire from active control of the properties. But in 1887 he once more got possession of the Northern Pacific with German capital and succeeded in arranging a lease of the Oregon Short Line, which had been developed by the Union Pacific interests, embracing a cross-country road from its main lines in Wyoming northward into Oregon and Washington. At the same time the interest of the Transcontinental Company in the Oregon Railway and Navigation Company was linked with the Oregon Short Line Company. These transactions, however, still left the Transcontinental Company in control of the situation, as it retained its majority ownership of Northern Pacific Railroad stock.

For the next few years the Northern Pacific did not follow a policy of rapid expansion. Other trunk lines, such as the Union Pacific, Rock Island, Santa Fé, Burlington, and North Western, were all growing and keeping pace with the rapid settlement of the West; but the Northern Pacific in these years simply rested content with its position as a single track transcontinental route having but few branches. Its only important extension was made by acquiring the Wisconsin Central Railroad, which gave the company a line between St. Paul and Chicago and a valuable and important entrance into the latter city. It was expected that with this accession, the affairs of the company would be permanently established on a sound basis, but the overliberal policy of paying out practically all the surplus in dividends was continued in the face of large increases in fixed charges.

Early in 1892 it began to be rumored that the Northern Pacific was not in so easy a financial position as had been assumed. The stockholders took alarm; and the committee which was appointed to investigate the situation discovered a deplorable state of affairs. As a result of the severe criticism of Villard's policy, steps were at once taken to oust him from control, but without success until June, 1893. Two months later, receivers were appointed who discovered that the company was insolvent and had no funds to pay quickly maturing obligations. Receivers were appointed also for most of the branch lines, including the Wisconsin Central system. The Oregon Short Line, which was tied through guarantees with the Union Pacific although leased to the Northern Pacific, was involved in the general crash but was later separately reorganized.

To rehabilitate the Northern Pacific Railroad effectively was a difficult problem. Its debt was enormous; its roadbed and rolling stock had been neglected; and, as a result of the recent crash, its valuable feeders on both east and west, the Wisconsin Central and the Oregon properties, were removed from its control. Besides these adverse conditions, competition of a serious nature was looming up. James J. Hill had for many years been quietly developing the Great Northern Railway. This great system he had financed in an extremely conservative manner; he had extended it through territory where construction costs were low; and he had secured control of branches and feeders which might have come under the sway of the Northern Pacific had that company been more farsighted. Hill had operated his road from the beginning at very low cost; he had kept its credit high; and even in the period of financial depression he had reported large profits and had paid substantial dividends on his stock. With such a competitor in the field, it really looked for a while as though the Northern Pacific could have no future whatever.

Finally, in May, 1895, a plan sponsored by Edward D. Adams, representing New York interests and those of the Deutsche Bank of Berlin, proposed a practical merger with the Great Northern Railroad Company: the old stock and bondholders were to make all the sacrifices and to supply all the new capital, and the Great Northern was then to be presented with half the stock of the new company, in consideration for which it was to guarantee the new Northern Pacific bonds. The situation was somewhat similar to that which existed in New York State as early as 1868 when Commodore Vanderbilt had achieved his great reputation as a wizard at railroading by acquiring the Harlem and Hudson River railroads and by forcing the New York Central lines to terms. James J. Hill had become a modern wizard, and the only hope for the Northern Pacific seemed to be to lay the road at his feet and ask him to do with it what he had done with the Great Northern — make it a "gold mine."

This plan, however, met with too much opposition and was abandoned. During the following year a new plan, backed by both the American and the German interests, secured the strong cooperation and endorsement of J. P. Morgan and Company. This was the first instance of Morgan's entry into railroad reorganization in the West. During the previous few years he had been increasing his reputation as a reorganizer of Eastern railroad properties, and by this time he had successfully organized or was rehabilitating the Erie, the

Reading, the Baltimore and Ohio, the Southern,
and the Hocking Valley systems. But he had kept
clear of the far Western field and had definitely
refused to reorganize the Union Pacific on the
ground that its territory was too sparsely settled
and that there was little hope for its future, especially as its partial control by the United States
Government made any reorganization extremely
difficult. The new plan for the Northern Pacific
was carried out with no regard to the Hill interests:
the old stockholders were heavily assessed; all
bondholders were forced to make sacrifices; the
Wisconsin Central lines were entirely eliminated
and separately reorganized; and the Oregon lines
were dissociated from the Northern Pacific and
afterwards returned to the control of the new
Union Pacific.

While the new Northern Pacific as reorganized
in 1898 came directly under Morgan's control and
was immediately classed as a Morgan property,
it did not remain exclusively such for very long.
In the promotion and development of the Great
Northern system, Hill had hitherto maintained an
independent position so far as banking alliances
were concerned, but he now began to develop
closer relations with the Morgans and became

heavily interested in the First National Bank of New York, an institution which for many years had been more or less directly identified with the Morgan interests. On more than one occasion thereafter the banking firm of J. P. Morgan and Company acted as financial agent for the Great Northern.

Soon after the reorganization of the Northern Pacific, it became known that Hill had acquired an important interest in the property, and as time went on this interest was substantially increased. Within a year or two the Northern Pacific began to be classed as one of the Hill lines. With a substantial Hill representation on the board of directors and a managerial policy which was clearly inspired by Hill, the company now entered upon a new stage in its career.

The outstanding dramatic event in the story of the modern Northern Pacific was the famous corner which occurred in the spring of 1901 as a result of a contest between the Hill and the Harriman interests for the control of the property. The details of this operation, which sent the price of Northern Pacific stock up to $1000 a share and precipitated a stock-market panic, form part of the story of the Harriman lines. The contest resulted in the formation of the Northern Securities Company, a corporation of $400,000,000 capital, devised as a holding company under the joint control of the Hill and Harriman interests, for the purpose of retaining a majority of the stocks of the Northern Pacific and the Great Northern.

The Hill interests, jointly with the Morgan control of the Northern Pacific, had been quietly accumulating stock in the Chicago, Burlington and Quincy Railroad, and Harriman felt that there was grave danger to the Union Pacific in this move, as the Burlington had already penetrated into the Union Pacific territory and might at any time start to build through to the coast its own line parallel to the Union Pacific. Harriman consequently began to buy up Northern Pacific stock in the open market and thus, together with the efforts of the Hill and Morgan people to retain and strengthen their control, brought about the corner.

The Northern Securities Company was designed to harmonize all interests and to keep the control of the Burlington property jointly in the hands of Harriman and Hill. But as the result of a suit under the Sherman Anti-Trust Act, this combination was declared illegal, and in 1904 the company was

dissolved. The final outcome of the situation was
that the Northern Pacific, sharing with the Great
Northern the joint control of the Burlington lines,
was left indisputably in the hands of the Hill-
Morgan group, where it has ever since remained.
These three great railroad systems, the Northern
Pacific, the Great Northern, and the Chicago, Burlington and Quincy, constituting nearly twenty
thousand miles of railroad, have been known ever
since as Hill lines.

Since the dramatic days of the Harriman-Hill
contest the history of the Northern Pacific system
has been simply a striking reflection of the growth
in population and wealth of the great Northwest.
The States through which it operates have grown
with astounding rapidity during the past two decades; small cities have spread into great centers of
manufacture and trade; hundreds of smaller towns
have sprung up; natural resources of untold value
have been developed. In the meanwhile the Northern Pacific has forged ahead in its earnings and
profits, and the stock of the road has come to be
known as one of the highest class of investment
issues. Although new competition appeared, in both
the local and the through business of the company
— notably by the extension of the St. Paul system

largely through Northern Pacific territory to the
Puget Sound region — the superior modern business management of James J. Hill, backed by the
strong resources of the Morgan banking interests,
made the Northern Pacific one of the standard
railroad systems of America.

CHAPTER VIII

THE Santa Fé Route, or the Atchison, Topeka and Santa Fé Railroad, which has in modern times developed into one of the largest and most profitable railroad systems in this country, was projected long before the idea of a transcontinental line to the Pacific coast had taken full possession of men's minds. As early as 1858 a plan was worked out for the construction of a line of about forty miles within the State of Kansas to connect what were then the obscure and unimportant townships of Atchison and Topeka. At that time not a mile of railroad had been built in Kansas or in any Territory west of that State, except on the Pacific coast, to which there had been an enormous immigration occasioned by the wonderful discovery of gold.

The outbreak of the Civil War delayed the undertaking of the Atchison-Topeka line, and

nothing more was done until 1863. In that year
new interests took control of the enterprise and
acquired rights for its extension through southwestern Kansas in the direction of Santa Fé, the
capital of the Territory of New Mexico. The company, which had originally been the Atchison and
Topeka, now changed its name to the Atchison,
Topeka and Santa Fé and obtained from the Government a very valuable land grant of 6400 acres
for every mile constructed, the only condition being
that within ten years the line should be completed
from Atchison to the western border of Kansas.
The plan involved the building of only 470 miles
of road, which when finished would assure the
company nearly three million acres of land within
the State of Kansas.

A decade would seem to be ample time for the
construction of this comparatively short railroad,
particularly with the inducement of so extraordinary a land grant. Not only the Union Pacific
but the Central Pacific and Kansas Pacific — all
built within this decade — had to accomplish far
more construction in order to secure their respec-
tive grants, and yet they had their complete lines
in operation years before the Santa Fé had fifty
miles of track in actual commission. The reason

for this delay was of course a financial one. The other roads had all received government aid in cash or securities in addition to land grants. But the Atchison line was, from the start, thrown on its own resources in raising capital, and it was not until late in 1869 — nearly a year after the opening of the Union Pacific to the coast — that any construction work whatever was done. In that year the section from Topeka to Burlingame, consisting of about twenty-eight miles, was opened for traffic, and a year later the extension to Emporia was finished, thus making a total of sixty-one miles under operation.

The terms of the land grant provided that the entire line across Kansas should be completed by June, 1873. When by 1872 only sixty-one miles of track had been built, the company still had over four hundred miles to go within ten months if it expected to obtain the land grant. But so energetically did the owners of the property work from that time on that within seven months they had reached the eastern boundary of Colorado and had thus saved the grant.

But like most of the Western railroads built in those early days the Santa Fé property was, in a sense, ahead of its time. The rapidity with which it shot across the State of Kansas in 1872 was equaled only by the promptness with which it fell into financial straits. No sooner had its complete line been opened for traffic than the panic of 1873 occurred; the company became embarrassed by a large floating debt; and a compromise had to be made with the bondholders whereby a postponement of a year's interest was arranged.

No attempts were made to extend the Santa Fé during the long period of depression following the panic of 1873. The road ended in 1872 at the Colorado state line, and during the next few years the only building of importance was a western spur to connect with the Denver and Rio Grande at Pueblo, thereby giving an outlet to the growing city of Denver and the rapidly developing mining regions of Colorado. About 1880, construction was resumed in a leisurely way, down the valley of the Rio Grande into New Mexico and in the direction of Albuquerque. In this extension, as in later building, the line of the old Arizona trail was usually followed. One writer has declared that "the original builders of the Atchison followed the line of the Arizona trail so religiously that if the trail skirted a ten-foot stream for a quarter of a mile to strike a shallow spot for fording, the railroad not supply traffic in sufficient amount even to "feed the engines."

To extend somewhere, then, was an imperative necessity. But whither? Several routes were under consideration. The Southern Pacific lines had worked eastward to El Paso on the Mexican border, several hundred miles due south from Albuquerque, and it looked feasible to extend the Atchison to that point and arrange a traffic agreement with the Southern Pacific, or to build an extension through New Mexico to Deming and then westward along the river valleys and down into Mexico to Guaymas on the Gulf of California. It was possible, in the third place, to build directly west from Albuquerque through Arizona and Southern California to the coast. Ultimately all of these plans were carried out.

The first extension of the Santa Fé was to Deming, New Mexico, where in March, 1881, its tracks met those of the Southern Pacific, and by agreement the company secured the use of the Southern Pacific to Benson, Arizona. From the first this new through route to the Pacific began to pay handsomely. Later on the line into Guaymas, Mexico, was added by the purchase of the Sonora Railway. Soon afterward the Santa Fé secured from the St. Louis and San Francisco Railway a half interest in the charter of the Atlantic and Pacific, a company which planned to build through to the coast. Meanwhile the St. Louis and San Francisco had been acquired by the Gould and Huntington interests, which, as the owners of the Texas and Pacific and the Southern Pacific systems, naturally opposed the plans of the Santa Fé. The matter was compromised by the agreement of the Santa Fé to build no farther west than the Colorado River, where the Santa Fé was to be met by an extension of the Southern Pacific line from Mojave, California.

This arrangement proved unprofitable to the Santa Fé, for the Southern Pacific naturally diverted traffic to El Paso and Ogden. A new arrangement was accordingly made in 1884, involving the purchase, by the Atlantic and Pacific, of the Southern Pacific division between Needles and Mojave, the obtaining of trackage rights between Mojave and San Francisco, and the use of the Southern Pacific terminals at San Francisco. To assure a connection with the coast in Southern California, the Santa Fé built a line to Colton, acquired the California Southern Railway from Colton to San Diego, and effected an entrance to

Los Angeles by leasing the Southern Pacific tracks from Colton.

The Santa Fé had now reached the Pacific coast over its own lines, but it was handicapped by poor connections with the East. Its next move therefore was eastward to Chicago, where it acquired the Chicago and St. Louis Railroad between Chicago and Streator, Illinois, and then constructed lines between the latter point and the Missouri River. During the same year the company opened branches southward to the Gulf of Mexico, until by May, 1888, the entire system comprised 7100 miles.

This rapid expansion of the property, combined with extravagance in management and a reckless policy in the payment of dividends, brought the company into financial difficulties within a year after the completion of the system. Unprofitable branches had been built, and these had become an immediate burden to the main system. It is the same story that has been told of most of the large railroads of those days. Strenuous efforts were made to save the property from a receivership, and a committee was appointed in September, 1889, to devise ways and means of reform and reorganization.

11

The new management of the Santa Fé was a rational one and substantially reduced the obligations of the road.   Had its spirit been maintained, a second failure and reorganization a few years later would not have been necessary.   New interests, however, came into the property, and, though it was hoped that they would support a conservative policy, the former programme of expansion was resumed until in 1890 the St. Louis and San Francisco system was merged with the Santa Fé on a very extravagant basis.   Within a year it was clear that the St. Louis and San Francisco would prove more of a liability than an asset. During the same time the less important purchase of the Colorado Midland Railway also turned out to be a poor investment.

The next four years were marked by more bad financial management which culminated in the failure of the reorganized company.   In 1892 an exchange of income bonds for fixed interest-bearing bonds so increased the fixed charges of the company that, as a result of the panic of 1893 and its ensuing depression, the great Santa Fé system suddenly found itself in the hands of a receiver. The president, John W. Reinhart, had persistently asserted throughout 1893 that the company was financially sound; but an examination of its books subsequently made in the interest of the security holders disclosed gross irregularities, dishonest management, and manipulation of the accounts.

During the year 1894 the property was operated under the protection of the courts, and early in 1895 a new and comprehensive scheme of reorganization was carried out. This latest plan involved dropping the St. Louis and San Francisco system, the Colorado Midland, and all other unprofitable branches; it wiped out the floating debt; it supplied millions of new capital; and it enabled the succeeding management at once to build up and improve the property.

At the head of the new company was placed Edward P. Ripley — a railroad manager of great executive ability and a practical, broad-minded business man of the modern type, who has ever since remained president of the road. The history of the Santa Fé since 1895 has been closely identified with Ripley's business career, and its record during these two decades has been an enviable one. Steady progress from year to year in volume of business, in general development of the system, in improvement of its rights of way, terminals, and equipment, has characterized its history

CHAPTER IX

THE GROWTH OF THE HILL LINES

THE States which form the northern border of the United States westward from the Great Lakes to the Pacific coast include an area several times larger than France and could contain ten Englands and still have room to spare. The distance from the head of the Great Lakes at Duluth to the Pacific coast in the State of Washington is greater than the distance from London to Petrograd or the distance from Paris to Constantinople, and three times the distance from Washington, D. C., to Chicago.

Fifty years ago these States, with the single exception of Wisconsin, were practically a wilderness in which only the Indian and buffalo gave evidences of life and activity. No railroads penetrated the forests or the mountain ranges. Far southward some progress in the march of civilization had been made; the Union Pacific had linked

the West with the East before the eighth decade of
the century began, and the Northern Pacific pro-
ect was being painfully pushed through the inter
mediate tier of States during the seventies. Bu
the material resources of the Great Northwest ha
still to be discovered.

When the Northern Pacific Railway failed i
1873, the crash involved a little railroad know
as the St. Paul and Pacific, running out of St. Pa
for a couple of hundred miles westward, with
branch to the north joining the Northern Pacifi
at Brainerd, Minnesota. The St. Paul and Pacifi
had been acquired in the interest of the Norther
Pacific some years earlier but was now regarde
as a property so worthless that its owners woul
be glad to get rid of it, if only they could find
purchaser rash enough to take it over.

During the three years following the panic o
1873 the crops of Minnesota were practically eat
en up by the grasshoppers, and poverty reigne
among the farmers. At that time a short, stock
man with long hair, one blind eye, and the reputa
tion of being the greatest talker in town, kept
coal and wood store in St. Paul. His name wa
James J. Hill. For years he had been a familia
figure, sitting in his old chair in front of his stor

nd discoursing on current events. This man was ot only an interesting talker; he was a visionary, dreamer — and one of his dreams was to buy the t. Paul and Pacific Railroad and to transform it ito a real railway line. Nearly twenty years ad passed since he had drifted in, an eighteen-ear-old Scotch-Irish boy from Ontario, and had egun work in a steamship office on the levee at t. Paul. Now, in 1876, he was thirty-eight years ld and a town character. And the town felt that , had his measure. He had already tried a vari-ty of occupations, and at this time was agent for nes of steamboats on the Mississippi and the ᷰed River. Everybody knew him and liked him, ut no one took him very seriously. The idea f his controlling the St. Paul and Pacific was ven amusing.

Now the most promising part of the St. Paul nd Pacific when it failed in 1873 was the line from t. Paul to Breckenridge on the Red River. Hill ᷰas the Mississippi steamboat agent at one end; t the other, an old Hudson Bay trader, Norman ᵂ. Kittson, ran two little old stern-wheel steam-oats from Breckenridge to Winnipeg. A large art of the freight that Hill and Kittson handled ᷰas for the Hudson's Bay Company. It came up

the Mississippi, went across on the St. Paul and
Pacific to Breckenridge, and then down the Red
River on Kittson's steamboats until it was re
ceived at Fort Garry, Winnipeg, by Donald Alex
ander Smith, then commissioner for the Hudson'
Bay Company.

Smith, who became afterwards Lord Strathcona
and High Commissioner for Canada in England
was a tall, lean, urbane Scotchman with a sof
manner and a long red beard. In 1876 he wa
fifty-six years old, with a life of strange, wild ad
venture behind him. He had gone when little mor
than a boy to Labrador to take charge of a statior
of the Hudson's Bay Company. Among the north
ern Indians he stayed for thirteen years. In th
sixties he was practically king over all the savag
territory of the company along the waters enter
ing Hudson Bay. By the seventies he was a mar
of means and he had some influence in the ne
Dominion of Canada.

It would be a great advantage to Smith to hav
a good railroad from St. Paul to Winnipeg as th
Red River boats were frozen up in the winter an
the service on the St. Paul and Pacific, under th
receiver, was impossible. So Smith listened wit
favor to Hill's project of getting hold of the St

Paul and Pacific and making a real railroad out
of it.   And whenever Smith went to Montreal he
talked the matter over with his cousin George
Stephen — later Lord Mount Stephen — who was
the head of the Bank of Montreal.   In 1877
Stephen and Richard B. Angus, the general manager of the Bank, went to Chicago on business.
While there, they had two weeks' time on their
hands, and tossed a penny to decide whether to
run down to St. Louis or up to St. Paul.   The
penny sent them to St. Paul.   "I am glad of that,"
said Stephen; "it will give us a chance to see the
prairies and look over that St. Paul and Pacific
road that Donald Smith is always talking about."

When they arrived in St. Paul, James J. Hill
took them over the line to Breckenridge.   The
country had been scoured by the grasshoppers and
looked like the top of an old rusty stove.   But
Stephen was a broad-minded man, wise enough to
know that the pest of grasshoppers could not last
forever.   He was greatly impressed with the ultimate possibilities of the soil and, under the hyp-
notic influence of Hill's eloquence, became quite
enthusiastic over the scheme for getting hold of
the railroad; but, as it would evidently involve
millions, he didn't see how it could be done.

The road had originally been financed by bond
sold largely in Holland, and to do anything at al
it was necessary to get in touch with these Dutc
bondholders. In 1877 Stephen went over to Am
sterdam and secured an option on the bonds a
thirty cents on the dollar — less than the accrue
interest which was due and unpaid on them. H
then came back to America, conferred with John
S. Kennedy at New York, who represented botl
Dutch and American bondholders, and brough
Kennedy into the combination.

In the spring of 1878 the St. Paul and Pacifi
was taken over. People still smiled at Hill and
wondered how he had induced a hard-headed banl
president like Stephen to put up the money. No
body in St. Paul believed in the future of the road
Even the syndicate's attorneys, when offered a
choice between taking $25,000 in cash or $500,000
of the new road's stock for their services, preferre
the cash. Had they taken the stock and held i
for thirty years, they would have had, in principa
and interest, some $30,000,000.

To the surprise of everybody, including Hill and
his friends, the grasshoppers suddenly disappeare
in the early summer of 1877 and never came back
That summer saw the biggest wheat crop tha

had ever been harvested in Minnesota. "Hill's Folly," as it was afterwards called, with its thirty locomotives and few hundred cars, was feverish with success. Hill worked every possible source to get extra cars and went all the way to New York to buy a lot of discarded passenger coaches from the Harlem Railroad. By the end of the season it was evident to everybody that the St. Paul and Pacific was going to have a career and that "Jim" Hill's dream was coming true.

Immediately the fortunate owners began to plan for the future. They had acquired the road at an initial cost of only $280,000 in cash. In the following year they advanced money for the completion of the unfinished section, as necessary to obtain the benefit of a generous grant of land from the State. Then, in 1879, having acquired full possession of the property, and having several millions of dollars in profits, they issued bonds for further developments. This gave them sufficient basis to enlarge their scheme greatly, and in the formation of the St. Paul, Minneapolis and Manitoba Railroad, they created $15,000,000 of stock, which was divided equitably among Hill, Stephen, Angus, Smith, Kennedy, and Kittson. This stock was all "water," but the railroad prospered so extraordinarily in the succeeding few years that by 1882 the stock was worth $140 a share. And in 1883 they issued to themselves $10,000,000 of six per cent bonds for $1,000,000 — a further division of $9,000,000, coming out of nothing but good will, earning power, and future prospects.

The decade from 1880 to 1890 witnessed a steady growth of the system formed in 1879 under the name of the St. Paul, Minneapolis and Manitoba. The 600 odd miles which it embraced when Hill and his coterie made their big stock division had grown in 1890 to 2775 miles. It then consisted of a main line reaching from St. Paul and Minneapolis across Minnesota and the northern part of North Dakota, far into Montana, with a second main line from Duluth across Minnesota to a junction with the St. Paul line in North Dakota, besides numerous branches reaching points of importance in both these States.

But the development of the Hill properties had by no means reached its limit at this time. Hill's dream had been to construct a through line across the northern tier of States and Territories to the Pacific, and this plan had been constantly in his mind while he was building up the system in Manitoba. The original line running up into Manitoba and reaching Winnipeg was all very well as a start. It had paid so well that the original group of men had become millionaires almost overnight. But Hill meant to show the public that, after all, the early success was only an incident and merely a stepping-stone to the really great thing.

Practical railroad men everywhere ridiculed the idea of a railroad running across the far northern country, climbing mountain ranges, traversing hundreds of streams and extending for great stretches through absolutely wild and uninhabited regions. Especially did they deem it absurd to attempt such an undertaking without government aid, subsidies, or grants of land, pointing to the experience of such roads as the Union Pacific, Northern Pacific, and Santa Fé. All these had received financial assistance and large land grants, and yet all had gone through long periods of financial vicissitude before they had become profitable and stable enterprises.

But Hill was more farseeing than his critics. In 1889 the name of the company was changed to the Great Northern Railway, and under this title the extension to the coast was rapidly carried forward and was opened in the panic year of 1893. When all the other transcontinental lines went into bankruptcy, Hill's road not only kept out of the courts but actually earned and paid annual dividends of five per cent on its stock. The five years from 1896 to 1901 were years of uninterrupted prosperity for the Great Northern Railroad. Each year its credit rose; each year it grew to be more of a force in the Western railway situation. In these years the control of the property had somewhat changed and a few of the original promoters had died or had withdrawn. But Hill, Lord Strathcona, Lord Mount Stephen, and John S. Kennedy of the original group, all held their large interests, and Hill in particular had added to his holdings as the years had gone by.

The secret of Hill's striking success with his Western extension was the method by which the line was constructed. Hill had a theory that it was far better to go around mountains and avoid grades than to climb them or to bore through them; it was always better to find the route which would make long hauls easy and economical. He thus built his road with the idea of keeping down the operating costs and of showing a larger margin of profit than the others. From the very start the Great Northern was noted for its low ratio of operating expenses and its comparatively long trains and heavy trainloads. It was by this method that it really made its money.

By the year 1901 the Great Northern Railway absolutely controlled its own territory. But it was still handicapped by lack of an independent entrance into Chicago, as its eastern lines terminated at Duluth and St. Paul. At the western end also, the situation was unsatisfactory. It seemed important for the Great Northern to control a line of its own into Portland, Oregon, because the Northern Pacific Railroad, which, as we have seen, had been reorganized several years before by the Morgan interests, had been rapidly extending its lines in Oregon and Washington. Hill and his associates, therefore, had been quietly buying a substantial interest in the Northern Pacific property and thus, in the course of time, had come into closer relations with the Morgan group in New York. Soon afterward, under Hill's influence, the Northern Pacific began the construction of further extensions in Oregon and reached into territory that the Harriman interests in the Union Pacific Railroad had regarded as their own. This move created much friction between the Harriman and Hill groups, and in order to forestall danger Harriman in turn began quietly accumulating an interest in the Northern Pacific property by purchases in the open market.

The story of the battle royal between the Hill and Harriman interests will be told in a subsequent chapter. It is not necessary to repeat the history of the famous corner of 1901 nor of the compromise effected by the formation of the Northern Securities Company. The final result of this contest was the complete harmonizing of the Western railroad situation, so far as the Hill and the Harriman interests were concerned. In the succeeding years the Great Northern system penetrated to the heart of Manitoba and constructed lines through British Columbia to Nelson and Vancouver. It built other branches to Spokane, Washington, and Helena and Butte, Montana. Moreover by the discovery of extensive ore deposits on the lines of the company in northern Minnesota and by subsequent purchases of other mines, the Great Northern acquired control of about sixty-five thousand acres and hundreds of millions of tons of iron ore. All the properties so controlled were leased on a very profitable basis to the United States Steel Corporation. The Great Northern Railroad itself did not retain control of the ore lands but, through a trusteeship, gave a beneficial interest in them to its stockholders in the shape of a special dividend.

The profits under this lease promised to be very large in the course of time, but the Steel Corporation had the option to cancel after a five-year period, and in 1912, as the result of a United States Government suit for the dissolution of the Steel Corporation, the lease was canceled. Since that time the trustees of the ore lands have executed other leases, and the Great Northern ore certificates are bringing in a substantial return to their owners.

The three Hill lines — the Great Northern, the Northern Pacific, and the Chicago, Burlington and Quincy — have been unusually profitable. The Great Northern and the Northern Pacific have steadily paid liberal dividends to their stockholders on increasing amounts of capital stock; and the Burlington, whose whole stock is owned by these two roads, has also handed over liberal profits year by year, at the same time accumulating an earned surplus of more than one hundred million dollars and spending an almost equal amount of profits on the improvement and maintenance of the property. The Burlington today controls the Colorado

12

Southern, which extends southward from the Burlington lines in Wyoming, passing through Denver, Pueblo, Fort Worth, and other points southward to the Gulf.

CHAPTER X

THE RAILROAD SYSTEM OF THE SOUTH

In the year 1856 a small single-track railroad was opened from Richmond to Danville, Virginia. This enterprise, like many others in ante-bellum days, was carried out largely with funds supplied by the State. As long afterwards as 1867, three-fifths of the stock was owned by the State of Virginia, but soon after this time the State disposed of its investment to a railroad company operating a line in North Carolina from Goldsboro westward to Greensboro, and projected southward to Charlotte. In modern times, this little road, like the Richmond and Danville, has become an integral part of the Southern Railway system, but in those days it was controlled, curiously enough, by the Pennsylvania Railroad Company.

After 1867 the new owners of the Richmond and Danville began aggressively to extend their lines. By leasing the North Carolina Railroad, a small

property forming a link with the Greensboro line,
they created a through route from Richmond to
Charlotte. By 1874 they had built the road southward to Atlanta, Georgia, and had thus formed
the first continuous route from Richmond to that
city. Because of the extreme disorder and depression in the South during the years after the
Civil War the line did not prosper and was sold
under foreclosure about 1875. But the company
was reorganized in 1878 and acquired the Charlotte, Columbia and Augusta, thus extending its
lines into the heart of South Carolina and tapping
a rich territory. During these early years the
Pennsylvania Railroad interests, which still held
control, supplied the funds necessary for making
improvements.

At the same time that the Richmond and Danville was linking up the commercial centers of
the southern Atlantic seaboard, another system —
known as the East Tennessee, Virginia and Georgia — was being built up in the Appalachian
Mountains to the west. This property and its
predecessors had to some extent been state-owned
enterprises at first, but in 1870 the Pennsylvania
Railroad interests acquired control. A holding
company called the Southern Railway Securities

Company was now formed for the purpose of controlling all the Pennsylvania Railroad interests south of Washington. Besides the properties mentioned, this Securities Company soon obtained several other Atlantic seaboard properties extending from Richmond to Charleston, and also the Memphis and Charleston Railroad, running from Memphis to Chattanooga.

Thus at this early day a considerable railroad system had been welded together in the South, reaching many points of importance and forming direct connection at Washington with the northern properties of the Pennsylvania system. Had this experiment been successful, we would perhaps today reckon the great Southern Railway system as part of the Pennsylvania group. But the outcome was disappointing; the roads did not prosper; and soon the poorer sections began to default. The Pennsylvania then disposed of its interests and left the roads to shift for themselves.

The East Tennessee was the best of these minor lines, and in 1877 it began to acquire others extending through the South. Soon it had penetrated the heart of Alabama, reaching what is today known as the Birmingham district. Additional extensions were made to Macon and Rome,

Georgia, and on the north an alliance was arranged with the Norfolk and Western, while with a view to securing some of the business of the West, a connection was constructed at Kentucky-Tennessee state line. Such was the condition of the East Tennessee property by the end of 1881. In the meantime the Richmond and Danville had practically stood still.

About this time a definite revival set in throughout the South as the long-drawn-out period of depression following the war came to an end. Railroad activity revived, and both the East Tennessee, Virginia and Georgia and the Richmond and Danville roads passed into the hands of new and more aggressive interests. The new owners constructed the Georgia Pacific, which ultimately stretched across Alabama and Mississippi. To finance this enterprise and to consolidate their interests, a new holding company — the Richmond and West Point Terminal Railway and Warehouse Company — was formed in 1881 with large powers and authority to acquire the stocks and bonds of railroad properties in many Southern States. In addition to the properties already named, the Virginia Midland Railway was now acquired, and by 1883 the entire system had been merged under this organization. The company also secured the control of a line of steamboats running from West Point, Virginia, to Baltimore, and made close traffic arrangements with the Clyde line of steamers running between New York and Philadelphia and all important Southern points.

The personality at the head of the Richmond and West Point Terminal Railway and Warehouse Company was Calvin S. Brice, a man who had become increasingly prominent in railway affairs in the Southern States. Brice was something of a genius at combination and by 1883 had linked together and solidified the various properties in a very efficient manner. Nevertheless the competitive conditions of the time, combined with the necessarily more or less crude and hazardous methods adopted in financing and capitalizing the enterprise, prevented the credit of the organization from reaching a sound and secure level. The Tennessee properties especially proved an encumbrance, and they were almost immediately threatened with bankruptcy. Brice therefore decided to reorganize these subsidiary lines, and a new company called the East Tennessee, Virginia and Georgia Railway took over this section of the system in 1886.

In the meanwhile the Richmond and Danville properties, which were themselves becoming burdened with an ever growing debt, gave the Brice interests constant trouble. A large amount of the stock of the Richmond and Danville, as well as most of its bond issues, remained still outstanding in the hands of the public. Consequently the only way in which Brice and his friends could save the Richmond and Danville property from completely breaking up was to merge it more closely with the holding company in some way. But the credit and standing of the holding company itself were anything but high, for in addition to paying no dividends it had piled up a heavy floating debt of its own and had a poor reputation in Wall Street.

The situation thus becoming acute, the management carried through a remarkable stock-juggling plan. Instead of merging the Richmond and Danville directly into the West Point Terminal Company, the directors secretly decided to turn the Terminal Company assets over to the Richmond and Danville without apprising the stockholders of the Terminal Company. In conformity with this plan, early in 1886 the Richmond and Danville leased the Virginia Midland, the Western North Carolina, and the Charlotte, Columbia and

Augusta railroads, and later in the year the Columbia and Greenville and certain other small lines. At about the same time the Richmond and Danville obtained in some unknown way large amounts of the Terminal Company stock, a portion of which it now issued in exchange for stocks and bonds of certain of these subsidiary companies which it had leased. Having carried through these transfers, the Richmond and Danville then threw the remainder of its Terminal Company stock on the market, where it was bought by investors who knew nothing about these secret transactions.

The Terminal Company was now left high and dry so far as the Richmond and Danville was concerned. But at this juncture a surprising thing happened. The management of the Terminal Company, in its turn, began to buy shares of Richmond and Danville stock and in a short time regained its former control. This shifting of power exactly reversed the situation which had previously existed, when the Terminal Company itself had been controlled by the Danville Company. These changes were followed by a further move on the part of the Brice and Thomas interests, which now formed a syndicate and turned over to the Terminal Company a majority of the stock of the East Tennessee Company for $4,000,000 in cash and a large amount of new Terminal Company stock.

When these transactions had been accomplished, the Terminal Company found itself once more securely in control of the entire system, and the Brice and Thomas interests had incidentally very considerably increased their fortunes and also their hold on the general situation. From this time, the Terminal Company went aggressively forward in an ambitious plan for further expansion. By acquiring control of the Central Railroad and Banking Company of Georgia, the Terminal management was involved with new financial interests which immediately sought to control the system and to eliminate the Brice and Thomas group. The consequent internal contest was adjusted, however, in May, 1888, by electing as president John H. Inman, a man who had been identified with the Central Railroad of Georgia system.

The Richmond Terminal system now put in motion further plans for expansion. In 1890 it acquired a system of lines extending south from Cincinnati to Vicksburg and Shreveport, known as the Queen and Crescent route, and in the meantime made a close alliance with the Atlantic Coast

Line system. By the end of 1891 the Richmond Terminal system embraced over 8500 miles of railroad, while the Louisville and Nashville, the next largest system in the Southern States, had only about 2400 miles.

But as 1891 opened, the vast Richmond Terminal system was perilously near financial collapse. Notwithstanding the great value of many of the lines, its physical condition was poor; the liabilities and capitalization were enormous; and much of the mileage was distinctly unprofitable. About this time many disquieting facts began to leak out: during the previous year the Richmond and Danville had been operated at a large loss, and this fact had been concealed by deceptive entries on the books; the dividends paid on the Central Railroad of Georgia stock had not been earned for some years; and the East Tennessee properties were hardly paying their way.

Various investigating committees were now appointed, and finally a committee headed by Frederic P. Olcott of New York took charge and worked out a complete plan of reorganization. The scheme, however, met with strenuous opposition, and thus matters dragged on into the panic period of 1893, when the entire system went into bankruptcy and into the hands of receivers. The various sections were operated separately or jointly by receivers during this unsettled period, and it looked for some time as though an effective reorganization which would prevent the properties from entirely disintegrating could not be successfully accomplished.

In the dark days of 1893, after Olcott and the Central Trust Company had failed to effect a reorganization of the Richmond Terminal system, a new interest came to the rescue, represented by the firm of J. P. Morgan and Company, whose growing reputation was due to the unusual personality of J. P. Morgan himself. He was essentially an organizer. The railroad properties which had become more or less identified with the Morgan interests had for the most part prospered. It was felt that Morgan's banking-house was the only one in Wall Street which might be equal to the task. The proposal was made to him; he did not invite it. In fact, it is said that for some time he was much opposed to taking hold of this disintegrated and broken-down system of railroads operating largely in poor and unprogressive sections, populated for the most part by negroes. Said Morgan, "Niggers are lazy, ignorant, and unprogressive; railroad traffic is created only by industrious, intelligent, and ambitious people."

After months of discussion, however, Morgan finally agreed to undertake the task, and out of the previous chaos there emerged the Southern Railway Company, which has been closely identified with Morgan's name ever since. Probably of the many railroad systems which Morgan reorganized from 1894 down to the time of his death, no system has become more distinctly a Morgan property than the Southern Railway Company.

The plan of reorganization whereby this great aggregation of loosely controlled and poorly managed Southern railroads was welded together into an efficient whole was a very drastic one in its effect on the old security holders. Debts were slashed down everywhere, assessments were levied, and old worthless stock issues were wiped out. Valueless sections of mileage were lopped off, and an effort was immediately made to strengthen those of real or promising value. Millions of dollars of new capital were spent in rebuilding the main lines; terminals of adequate scope were constructed in all centers of population; and alliances were made with connecting links with a view to building up through traffic from the North and the West.

The first ten years of the Southern Railway system under the Morgan control were practically years of rebuilding and construction. While after ten years of work the main system still radiated through most of the territory already occupied in a crude way in 1894, yet it had acquired a large number of feeders and smaller railroads in other sections. The Mobile and Ohio, operating with its branches about one thousand miles from Mobile to St. Louis, Missouri; the Georgia Southern and Florida, furnishing an important connection from the main system to various points in the State of Florida; the Alabama Great Southern, operating in and near the Birmingham district of Alabama — all these properties were molded into the system during these years. The system was then rounded out toward the North and consolidated through joint control, with the Louisville and Nashville, of the Chicago, Indianapolis and Louisville Railroad, which operated lines northward into Ohio and Illinois and on to Chicago. Thus, with the lines of the Queen and Crescent route running southward from Cincinnati to New Orleans, the system secured a direct through line from its various southern points to the shores of the Great Lakes.

In addition to these developments, the management of the Southern Railway system arranged direct connection with Washington through the joint acquisition with other lines of the Richmond, Fredericksburg and Potomac; it made traffic arrangements with the Pennsylvania and the Baltimore and Ohio systems to Baltimore, Philadelphia, and New York; and it also developed close alliances with the coastwise steamships plying northward from various Southern points.

In the reorganization of 1894 the Central of Georgia Railway system was cut off and separately reorganized, although it remained under the control of Morgan for a number of years. Finally in 1907 Morgan sold his Georgia properties to Charles W. Morse. They subsequently passed to Edward H. Harriman, who afterwards merged them into the Illinois Central system, under which control they have since remained.

As compared with the old Richmond Terminal aggregation with its broken-down rails and roadbed, poor equipment, and miserable service, the modern Southern Railway system shows startling changes. The Southern States have grown enormously in population and wealth during the last generation; the industrial activities of the South at the present time are elements of large importance to the country as a whole. Cities have vastly increased in population; new towns and manufacturing districts have been built up; and at the present there is scarcely a mile of unprofitable railroad in the entire 9000 miles under operation. In recent years large soft coal deposits have been discovered and developed on many of the branch lines, and today the coal tonnage of the Southern Railway is exceeding the relatively unstable lumber tonnage of two or three decades ago.

CHAPTER XI

THE LIFE WORK OF EDWARD H. HARRIMAN

In a previous chapter there has been related the early history of the great line that first joined the Atlantic and the Pacific Oceans — the Union Pacific. But the history of this property in recent years is almost as startling and romantic as its story in the sixties and seventies. It was not until recent days that the golden dreams entertained by these early builders came true. The man who really reaped the harvest and who at the same time gave the Union Pacific that position among American railroads which its founders foresaw was the last, and some writers think, the greatest of all American railroad leaders.

The Union Pacific, a bankrupt railroad in 1893, lay quiescent under the stress of the hard times that lasted until 1898. The long story of its tribulations hardly made it a tempting morsel for the men who were then most active in the railroad field.

In 1895 or 1896 the several protective committees which had been appointed to look after the interests of stockholders and defaulted bondholders had tried to induce J. P. Morgan to undertake the reorganization, but he had refused. To reorganize the Union Pacific meant that not far from one hundred millions of new capital would sooner or later have to be supplied, and there was no other banking-house in America at that time which seemed strong enough for the task. Smaller concerns were all involved in the Morgan syndicates or in other undertakings, and a combination of these at the moment seemed out of the question.

About this time the German-Jewish banking-house of Kuhn, Loeb and Company began looking into the situation. Kuhn, Loeb and Company were known as a very conservative but very rich concern with close connections in Frankfort and Berlin. Though it had been long established in New York it had not been identified with the railroad reorganization movement nor had it been prominent as an investing or underwriting institution. But now the active partner of the business, Jacob H. Schiff, set out seriously to persuade the various committees to adopt a plan of reorganization which he had devised. Though he made some progress, he soon found much secret opposition and thought that Morgan might be quietly attempting to secure the property. Morgan, however, was not interested. The mystery was still unsolved.

The fact was that Edward H. Harriman, who for some years past had been a powerful influence in the affairs of the Illinois Central Railroad but who was unknown to the average Wall Street promoter and totally unheard of throughout the country, had made up his mind to reorganize the Union Pacific Railroad. He therefore began to work quietly with various interests in an attempt to tie up the property. But soon he, like Schiff, encountered serious opposition. He also immediately jumped to the conclusion that Morgan was secretly at work, and he called on Morgan for the facts. Morgan replied, as he had replied to Schiff, that he was not interested, but that he wished Harriman success.

As Schiff continued to meet with difficulty, he soon called on Morgan again. Again Morgan replied that he was not interested. "But," he said, "I think if you will go and see a chap named E. H. Harriman you may find out something."

Who was Harriman? Schiff had hardly heard of him and had never met him. How could a small man like Harriman, with no money, no powerful friends, no big financial backing, reorganize a great system like the Union Pacific Railroad? The idea seemed ridiculous. Nevertheless, as the opposition continued, Schiff soon got in touch with Harriman. In the course of a conference, he warned this daring interloper to keep his hands off the Union Pacific. But Harriman was not moved by threats On the contrary, he insisted that Schiff should leave the Union Pacific alone; that he himself had already worked out his plans to reorganize it. Schiff laughed at this idea, termed it chimerical, and asserted that Kuhn, Loeb and Company were easily able to obtain the needed one hundred millions or more through their foreign connections on a basis of from four to five per cent, and that in America no such sum of new capital could at that time be raised through banking activities at better than six or seven per cent.

Harriman then sprang his surprise on Schiff. For some years he had been financially interested in the affairs of the Illinois Central. This property had at that time higher credit than any other American railroad; it had raised large sums of capital in Europe on as low a basis as three per cent, and on most of its bonds paid only three and one-half per cent interest. For nearly fifty years the property had been paying dividends with hardly an interruption, and altogether it had an enviable reputation as one of the soundest investments. Harriman's influence in the affairs of the company had been increasing quietly for years; the management had been left almost completely in his hands; and the directors were in effect largely his puppets, and a majority would do his bidding in almost anything he might propose.

Harriman now announced to Schiff that he intended to have the Union Pacific reorganized as an appendage of the Illinois Central. The necessary one hundred millions would be raised by a first mortgage on the entire Union Pacific lines at three per cent, and the mortgage would be guaranteed by the Illinois Central, while the latter company would receive a majority of the new Union Pacific stock in consideration for giving its guarantee.

Here was a poser for Schiff, who saw at once that if Harriman could use the Illinois Central credit in this way, he certainly could carry out his plan. Schiff soon found that Harriman would have no difficulty in using Illinois Central credit. The upshot of the matter was that the two men got together and jointly reorganized the Union

Pacific. Harriman was made chairman of the Board of Directors, and Kuhn, Loeb and Company became the permanent bankers for the new railroad system.

Thus with one bound Harriman had leaped to the forefront in American railroad finance and by a bold act which was characteristic of the man. For Edward H. Harriman was not only a hard-headed, practical business builder who like Morgan thought in big figures, but he was also a bold plunger, which Morgan was not. Possessing a vivid imagination, he not only saw far into the future but he also planned far into that same future. Morgan was also a man of vision, but his vision did not carry him far beyond the present. The things Morgan saw best were those immediately before him, while the things that Harriman saw best were at a distance. Morgan's big plans of procedure were based on what he saw in a business way in the near future; he reorganized his railroads with the idea of making them pay their way as soon as possible and of showing a good return on the capital invested. He thought little of what might be the outcome a decade or two hence or of what combinations might later be worked on the chessboard as a result of his immediate moves.

Morgan's mind was not philosophical; it was intensely practical.

While Morgan declined the proffered control of the Union Pacific on the theory that it was only a "streak of rust" running through a sparsely settled country and across an arid desert, Harriman dreamed of the great undeveloped West filling up with people during the following generation, of the empty plains being everywhere put under cultivation, and of the arid desert responding to the effects of irrigation on a large and comprehensive scale. He foresaw the wonderful future of the Pacific States — the opening up of natural resources in the mountains, the steady stream of men and women who would ultimately migrate to this vast section from the East and from foreign lands and who would build up towns and great cities. At the same time, with that practical mind of his, Harriman calculated that the Union Pacific Railroad — situated in the heart of this huge area, having the most direct and shortest line to the Pacific, and with all traffic from the East converging over half a dozen feeder lines to Omaha and Kansas City — would haul enormous amounts of tonnage just as soon as the Western country revived from the depression

under which it had been struggling for half a
dozen years.

When Harriman took hold of the Union Pacific
he had already determined to absorb the Oregon
lines, with their tributaries running up into the
Puget Sound country and to the Butte mining
district; to get hold of the Southern Pacific properties at the earliest possible moment; and to link
the Illinois Central in some way to the Union Paci
fic so that the latter would have its own independent outlets to Chicago and St. Louis. All these
plans he ultimately accomplished, as well as many
others, some of which his farseeing imagination
may have conceived then.

While Harriman was able very promptly to
carry through his first scheme and recapture the
Oregon lines, which had been separately reorgan
ized as a result of the receivership, he found it a
far more difficult matter to secure a dominating
interest in the great system of railroads controlled
by Collis P. Huntington. Huntington was a hard
man to deal with. Himself one of the practical
railroad magnates of his time, he also had the gift
of vision and undoubtedly foresaw that the ulti
mate result must be a consolidation of the properties; but he fully expected that his company

would absorb the Union Pacific. Had it not been that during the panic period the Southern Pacific had heavy loads of its own to carry and that its credit was none too high, Huntington might then have attempted to gain control of the Union Pacific.

Events finally worked to the benefit of Harriman. When Collis P. Huntington died in 1900, it was in most people's minds only a question of time as to when the powerful Harriman interests would take over the Southern Pacific properties. Consequently there was no surprise when in 1901 announcement was made that the Union Pacific had purchased the holdings of the Huntington estate in the Southern Pacific Company and was therefore in virtual control.

By a master stroke the railroad situation in the West had been radically changed. The Huntington system comprehended many properties of large and growing value, which were now feeling the full benefit of the agricultural prosperity at that time spreading throughout the great Southwest. Aside from this prize, the Union Pacific acquired the main line to the Pacific coast which it had always coveted and thus added to its system over nine thousand miles of railroad and over four thousand miles of water lines, besides obtaining a grip on the railroad empire of this entire portion of the continent not to be readily loosened by competitors.

At the same time that Harriman was strengthening his position on the west and south, the Great Northern and Northern Pacific properties, both now operated under the definite control of James J. Hill, were following a policy of expansion fully as gigantic as that of the Union Pacific. The Great Northern lines operating from Duluth to the Pacific coast had become powerful elements in the Western railroad situation, and Hill had devised many plans for diverting to the north the through traffic coming from the central section of the continent. He had established on the Great Lakes a line of steamships running from Duluth to Buffalo, and was also operating on the Pacific Ocean steamship lines which gave him a connection with Japan, China, and other oriental countries.

After the reorganization of the Northern Pacific Railroad, which fell under the domination of Morgan, the affiliations of the Hill and Morgan interests became very close, and in a short time Hill had as secure a grip on the Northern Pacific as he had always had on the Great Northern. This powerful combination looked like a menace to the Harriman-Kuhn-Loeb interests which controlled the territory to the south and radiated throughout the State of Oregon. When, therefore, the Northern Pacific began a little later to build into territory in Oregon and Washington which the Union Pacific regarded as a part of its own preserves, much bad feeling was engendered between the two interests. Matters were brought to a climax in the spring of 1901 when the Harriman people suddenly made the discovery that the Hill-Morgan combination had been quietly buying control of the valuable Chicago, Burlington and Quincy Railroad, which operated a vast system west and northwest of Chicago, penetrated as far into the Union Pacific main-line territory as Denver, and connected at the north with the eastern terminals of both the Great Northern and Northern Pacific systems. This move meant but one thing to Harriman: the Hill-Morgan interests were trying to surround the Union Pacific and make it powerless, just as the Southern Pacific had attempted to do many years before.

Harriman now played one of his bold strokes. He immediately began to purchase Northern Pacific stock in the open market in order to secure control of that property. It was well known that while the Hill-Morgan alliance dominated the

Northern Pacific, it did not actually own a majority of the stock, and to secure this majority was Harriman's purpose. This move would effectually check the invasion of the Union Pacific territory by giving the Harriman interests a voice in the control of the Chicago, Burlington and Quincy.

The price of Northern Pacific common stock soared day after day until on May 9, 1901, it sold at $1000 a share, and a momentary panic ensued. At the time Morgan was on the ocean and could not be reached. His partners were apparently not equal to the emergency. But Harriman was. When the panic reached its height, both interests had purchased far more than a majority of Northern Pacific stock — in contracts for future delivery. It was seen that to insist on the delivery of shares which did not exist would not only bankrupt every "short" speculator, large and small, but would undoubtedly bring all Wall Street tumbling down like a house of cards. So, in the midst of the excitement, the two interests reached a compromise.

The outcome was the formation of the Northern Securities Company with a capital of $400,000,000, nearly all of which was issued to acquire the capital stocks of the Northern Pacific and Great

Northern railroads. All the properties, including the Burlington, thus came under the joint control of the Harriman and Hill groups. The division of territory on both the east and the west was worked out amicably: the Northern Pacific abandoned some of its plans for extensions in Oregon, and the Burlington system remained as it was, with the understanding that no extensions should be built to the Pacific coast. Later the Burlington acquired control of a cross-country system, the Colorado Southern, extending south to the Gulf, but to this day has made no attempt to build beyond the lines it owned to Wyoming in 1901.

As is well known, the Northern Securities Company was subsequently declared to exist in violation of the Sherman Anti-Trust Act, and on a decision of the United States Supreme Court in 1904 it was practically dissolved and all its securities were returned to the original holders. This dissolution left the Hill-Morgan interests in undisputed control of the Burlington properties, but harmonious relations had in the meantime been established among the contestants, assuring an equitable division of territory and traffic. The final outcome was that the Union Pacific Railroad Company, which had purchased with its large and Ohio, a large block of whose stock was disposed of by the Pennsylvania Railroad. Harriman had already largely added to the Union Pacific's holdings in the Illinois Central. Jointly with the Lake Shore of the Vanderbilt system, the Baltimore and Ohio had, as already described, acquired a dominating interest in the Reading Company, including all the latter company's interests and affiliations as well as its entry into the New York district through control of the Central Railroad of New Jersey. Harriman, therefore, by a single stroke, now found himself in practical possession of a coast-to-coast system of railroads extending all the way from New York to San Francisco, Portland, and Los Angeles, and passing through all the important cities of the country. The Illinois Central system, operating nearly five thousand miles of road southward from Chicago to New Orleans, passing through St. Louis, with an arm reaching out to Sioux City on the west and a network of branches covering the Middle States, had thus become the great link welding together the eastern and western Harriman systems.

Later the Union Pacific acquired large interests in other properties and purchased substantial amounts of stock in the Atchison, Topeka and

Santa Fé, the New York Central. the St. Paul,
and the Chicago and North Western railroads.   It
also acquired a dominating interest in the Chicago
and Alton property, operating from Chicago to
St. Louis, with Western branches.   In the panic
period of 1907, Harriman personally purchased
from Charles W. Morse, who had acquired the
property from Morgan a short time before, the
entire capital stock of the Central of Georgia Railway, which he later turned over to the Illinois
Central.   The Central of Georgia lines connect
at several points with the Illinois Central and have
given the system various outlets on the South
Atlantic seaboard.

Harriman died in September of 1909, and with
his death the wizard touch was clearly gone.   What
would have been the later history of the Union
Pacific had he lived can be only conjectured.   The
new management, with Judge Robert S. Lovett
at its head, continued the broad and efficient operation which had characterized Mr. Harriman's régime, but it soon abandoned the policy of further
growth and expansion.   This alteration in policy,
however, was perhaps more the result of changing
conditions than of relinquishment of Harriman's
aims.   Many new laws for the regulation of the

railways had been passed, and in 1906 the powers of the Interstate Commerce Commission were greatly augmented. A period of reform had now begun, and after 1909 a wave of "progressivism" overspread the country. New interpretations were given to the Sherman Act, and suits were soon under way against all the railroads and industrial combinations which appeared to be infringing that statute. The great Standard Oil and Tobacco trusts were dissolved in this period, and a suit which was brought to divorce the Union Pacific and the Southern Pacific Company was finally decided against the Union Pacific, with the result that the two big properties were separated. The Union Pacific turned a large amount of its Southern Pacific stock holdings over to the Pennsylvania Railroad, in exchange for which it received from the Pennsylvania the remainder of the Baltimore and Ohio stock which the Pennsylvania interests had retained after the sale to the Union Pacific in 1906. Immediately after this, the Union Pacific management, seeing no particular advantage in retaining an interest in the Baltimore and Ohio, gave the shares to its own stockholders in a special dividend.

Thus, since Harriman's death, the Union Pacific

14

CHAPTER XII

THE AMERICAN RAILROAD PROBLEM

DURING the last fifty years the railroad has perhaps been most familiar to the American people as a "problem." As a problem it has figured constantly in politics and has held an important position in many political campaigns. The details that comprise this problem have been indicated to some extent in the preceding pages — the speculative character of much railroad building, the rascality of some railroad promoters, the corrupting influence which the railroad has too frequently exerted in legislatures and even in the courts. The attempts to subject this new "monster" to government regulation and control have furnished many of the liveliest legislative and judicial battles in American history. Farmers, merchants, manufacturers, and the traveling public have all had their troubles with the transportation lines, and the difficulties to which these struggles have given rise have produced that problem which is even now apparently far from solution.

Railroads had been operating for many years in this country before it dawned upon the farmers that this great improvement, which many had hailed as his greatest friend, might be his greatest enemy. It had been operating for several decades in the manufacturing sections before the enterprising industrialist discovered that the railroad might not only build up his business but also destroy it. From these discoveries arose all those discordant cries of "extortion," "rebate," "competition," "long haul and short haul," "regulation," and "government ownership," which have given railroad literature a vocabulary all its own and have written new chapters in the science of economics. The storm center of all this agitation concerned primarily one thing — the amount which the railroad might fairly charge for transporting passengers and freight. The battle of the people with the railroads for fifty years has been the "battle of the rate." This has taken mainly two forms, the agrarian agitation of the West against transportation charges, and the fight of the manufacturing centers, mainly in the East, against discriminations. Perhaps its most characteristic episodes have been the fight of the "Grangers" and their successors against the trunk lines and that of the general public against the Standard Oil Company.

Even in the fifties and the sixties, the American public had its railroad problem, but it was quite different in character from the one with which we have since grown so familiar. The problem in this earlier period was merely that of getting more railroads. The farmer pioneers in those days were not demanding lower rates, better service, and no discrimination and anti-pooling clauses; they asked for the building of more lines upon practically any terms. This insistence on railroad construction in the sixties explains to a great extent the difficulties subsequently encountered. In a large number of cases railroad building became a purely speculative enterprise; the capitalists who engaged in this business had no interest in transportation but were seeking merely to make their fortunes out of constructing the lines. Not infrequently the farmers themselves furnished a considerable amount of money, expecting to obtain not only personal dividends on the investment but larger general dividends in the shape of cheap transportation rates and the development of the country. Even when the builders were more honest, their mistaken enthusiasm had consequences which were similarly disastrous. The simple fact is that a considerable part of the Mississippi Valley, five or ten years after the Civil War, found itself in the possession of railroads far in excess of the public need. In the long run this state of affairs was probably not a great economic evil, for it stimulated development on a tremendous scale; but its temporary effect was disastrous not only to the railroads themselves but to the struggling population. The farmer had mortgaged his farm to buy stock in the road; and his town or county or State had subsidized the line by borrowing money which it frequently could not repay. When this property became bankrupt, not only wiping out these investments but leaving the agricultural population at the mercy of what it regarded as exorbitant rates and all kinds of unfair discriminations with high interest charges on its mortgages and high local taxes, the blind fury that resulted among the farmers was not unnatural.

Many of the railroad evils were inherent in the situation; they were explained by the fact that both managers and public were dealing with a new agency whose laws they did not completely understand. But the mere play of personal forces in themselves aggravated the antagonism. The fact that most of the railroad magnates lived in the East added that element of absentee landlordism which is essential to most agrarian problems. Many of the Western capitalists were real leaders; yet it is only necessary to remember that the most active man in Western railroads in the seventies was Jay Gould, to understand the suspicion in which the railroad promoter of that day was generally held. It is significant that of all the existing railroad abuses, the one which seemed to arouse particular hostility was the free pass. There were many greater practical evils than this, yet the fact that most editors and public officials and politicians and legislators and even many judges rode "deadhead" was a constant reminder of the influence which this "alien" power exercised over the government and the public opinion of the communities of which it was theoretically the servant. Many of these roads had a greater income than the States they served; their payrolls were much larger; their head officials received higher salaries than governors and presidents. The extent to which these roads controlled legislatures and, as it seemed at times, even the courts themselves, alarmed the people. The stock-jobbing that had formed so large a part of their history added nothing to their popularity.

Yet, when all these charges against the railroads are admitted, the fundamental difficulty was one which, at that stage of public enlightenment, was beyond the power of individuals to control. Nearly all the deep-seated evils arose from the fact that the railroads were attempting to do something which, in the nature of the case, they were entirely unfitted to do — that is, compete against one another. When the great trunk lines were constructed, the idea that competition was the life of trade held sway in America, and the popular impression prevailed that this rule would apply to railroads as well as to other forms of business. To the few farseeing prophets who predicted the difficulties which subsequently materialized, the answer was always made that competition would protect the public from extortion and other abuses. But competition between railroads is well-nigh impossible. Only in case different companies operated their cars upon the same roadbed — something which, in the earliest days, they actually did on certain lines — could they compete, and any such system as a general practice is clearly impracticable. One railroad which paralleled another in all its details might compete with it, but there are almost no routes that can furnish business enough for two such lines, and the carrying out of such an idea involves a waste of capital on an enormous scale. Probably the country received its most striking illustration of this when the West Shore Railroad in New York State was built almost completely duplicating the New York Central, with the result that both roads were nearly bankrupted.

While no one railroad can completely duplicate another line, two or more may compete at particular points. By 1870 this contingency had produced what was regarded as the greatest abuse of the time — the familiar problem of "long and short haul." Two or more railroads, starting at an identical point, would each pursue a separate course for several hundred miles and then suddenly come together again at another large city. The result was that they competed at terminals, but that each existed as an independent monopoly at intermediate points. The scramble for business would thus cause the roads to cut rates furiously at terminals; but since there was no competition at the intervening places the rates at these points were kept up, and sometimes, it was charged, were raised in order to compensate for losses at the terminals. Thus resulted that anomaly which strikes so strangely the investigator of the railroad problem — that rates apparently have no relation to the distance covered, and that the charge for hauling a load for seventy-five miles may be actually higher than that for hauling the same load one hundred or one hundred and fifty miles. The expert, looking back upon nearly a hundred years of railroad history, may now satisfactorily explain this curious circumstance; but it is not surprising that the farmer of the early seventies, overburdened with debt and burning his own corn for fuel because he could not pay the freight exacted for hauling it to market, saw in the system only an attempt to plunder. Yet even the shippers at terminal points had their grievances, for the competition at these points became so savage and so ruinous that the roads soon entered into agreements fixing rates or formed "pools." In accordance with this latter arrangement, all business was put into a common pot, as the natural property of the roads constituting the pool; it was then allotted to different lines according to a percentage agreement, and the profits were divided accordingly. As the purpose of rate agreements and pools was to stop competition and to keep up prices, it is hardly surprising that they were not popular in the communities which they affected. The circumstance that, after solemnly entering into pools, the allied roads would frequently violate their agreements and cut rates surreptitiously merely added to the general confusion.

The early seventies were not a time of great prosperity in the newly opened West, and the farmers, looking about for the source of their discomforts, not unnaturally fixed upon the railroads. Their period of discontent coincided with what will always be known in American history as "the Granger movement." In its origin this organization apparently had no relation to the dissatisfaction which its leaders afterward so successfully capitalized. Its founder, Oliver Hudson Kelley, at the time when he started the fraternity was not even a farmer but a clerk in the Agricultural Bureau at Washington. Afterward, when the Grangers had become an agrarian force to be feared, if not respected, it was a popular jest to refer to the originators of this great farmers' organization as "one fruit grower and six government clerks." Kelley's first conception seems to have been to organize the farmers of the nation into a kind of Masonic order. The Patrons of Husbandry, which was the official title of his society, was a secret organization, with signs, grips, passwords, oaths, degrees, and all the other impressive paraphernalia of its prototype. Its officers were called Master, Lecturer, and Treasurer and Secretary; its subordinate degrees for men were Laborer, Cultivator, Harvester, and Husbandman; for women — and women took an important part in the movement — were Maid, Shepherdess, Gleaner, and Matron, while there were higher orders for those especially ambitious and influential, such as Pomona (Hope), Demeter (Faith), and Flora (Charity). Certainly these titles suggest peace and quiet rather than discontent and political agitation; and, indeed, the organization, as evolved in Kelley's brain, aimed at nothing more startling than the social, intellectual, and economic improvement of the agricultural classes. Its constitution especially excluded politics and religion as not being appropriate fields of activity. It did propose certain forms of business coöperation, such as the common purchase of supplies, the marketing of products, perhaps the manufacture of agricultural implements; but its main idea was to contribute to the social well-being of the farmers and their families by frequent meetings and entertainments, and to improve farming methods by collecting agricultural statistics and by spreading the earliest applications of science to agriculture. The idea that the "Grange," as the organization was generally known, would ultimately devote the larger part of its energies to fighting the railroads apparently never entered the minds of its founders.

Had it not been for the increasing agricultural discontent against railroads and corporations in general, the Patrons of Husbandry would probably have died a painless death. But in the early seventies this hostility broke out in the form of minority political parties, the principal plank in whose platform was the regulation of the railroads. Farmers' tickets, anti-monopoly parties, and anti-railroad candidates began to appear in county and even state elections, sometimes achieving such success as to frighten the leaders of the established organizations. The chief aim of the discontented was "protection from the intolerable wrongs now inflicted on us by the railroads." "Railroad steals," "railroad pirates," "Wall Street stock-jobbers," and like phrases supplied the favorite slogans of the spirited rural campaigns. These parties, though much ridiculed by the metropolitan press, started a political agitation which spread with increasing force in the next forty years and in recent times eventually gained the ascendency in both the old political parties.

The panic of 1873 and the unusually hard times that followed added fuel to the flame. It was about this time that the Patrons of Husbandry gave evidences of a new vitality, chiefly manifested in a rapidly increasing membership. On May 19, 1873, there were 3360 Granges in the United States, while nineteen months later, on January 1, 1875, there were 21,697, with a total membership of over seven hundred thousand. In the Eastern States the movement had made little progress; in the South it had become somewhat more popular; in such States as Missouri, Iowa, Kansas, Nebraska, Montana, Idaho, and Oregon, it had developed into almost a dominating influence. It is not difficult to explain this sudden and astonishing growth: the farmers in the great grain States seized upon this organization as the most available agency for remedying their wrongs and rescuing them from poverty. In their minds the National Grange now became the one means through which they could obtain that which they most desired — cheaper transportation. Not only did its membership show great increase, but money from dues now filled the treasury to overflowing. At the same time the

organs of the capitalist press began to attack the
Grange violently, while the politicians in the sections where it was strongest sedulously cultivated
it. But the leaders of the movement never made
the fatal mistake of converting their organization
into a political party. It held no political conventions, named no candidates for office, and even
officially warned its members against discussing
political questions at their meetings. Yet, according to a statement in the *New York Tribune*,
"within a few weeks the Grange menaced the political equilibrium of the most steadfast States. It
had upset the calculations of veteran campaigners,
and put the professional office-seekers to more
embarrassment than even the Back Pay." The
Grangers fixed their eyes, not upon men or upon
parties, but upon measures. They developed the
habit of questioning candidates for office concerning their attitude on pending legislation and of
publishing their replies. Another favorite device
was to hold Granger conventions in state capitals
while the legislature was sitting and thus to bring
personal pressure in the interest of their favorite
bills. This method of suasion is an extremely
potent political force and explains the fact that,
in certain States where the Granges were most

powerful, they had practically everything their own way in railroad legislation.

The measures which they thus forced upon the statute books and which represented the first comprehensive attempt to regulate railroads have always been known as the "Granger Laws." These differed in severity in different States, but in the main their outlines were the same. Practically all the Granger legislatures prohibited free passes to members of the legislatures and to public officials. A law fixing the rate of passenger fares — the maximum ranging all the way from two and one-half to five cents a mile — was a regular feature of the Granger programme. Attempts were made to end the "long and short haul" abuse by passing acts which prohibited any road from charging more for the short distance than for the long one. More drastic still were the laws passed by Iowa in 1874 and the famous Potter bill passed by Wisconsin in the same year. Both these measures, besides fixing passenger fares, wrote in the law itself detailed schedules of freight rates. The Iowa act included a provision establishing a fund of $10,000 which was to be used by private individuals to pay the expenses of suits for damages under the act, and this same act made all railroad officials and employees who were convicted of violations subject to fine and imprisonment. The Potter act was even more severe. It not only fixed maximum freight rates, but it established classifications of its own. The railroads asserted that the framers of this law had simply taken the lowest rates in force everywhere and reduced them twenty-five per cent. But Iowa and Wisconsin and practically all the States that passed the Granger laws also established railroad commissions. For the most part these commissions followed the model of that established by Massachusetts in 1869, a body which had little mandatory authority to fix rates or determine service, but which depended upon persuasion, arbitration, and, above all, publicity, to accomplish the desired ends. The Massachusetts commission, largely owing to the high character and ability of its membership — Charles Francis Adams serving as chairman for many years — had worked admirably. In the most part these new Western commissions were limited in their activities to regulating accounting, obtaining detailed reports, collecting statistics, and enforcing the new railroad laws.

These measures, following one another in rapid succession, produced a national, even an international sensation. The railroad managements stood aghast at what they regarded as demagogic invasions of their rights, and the more conservative elements of the American public looked upon them as a violent attack upon property. Up to this time there had been little general understanding of the nature of railroad property. In the minds of most people a railroad was a business, precisely like any other business, and the modern notion that it was "affected with a public interest" and that the public was therefore necessarily a partner in the railroad business had made practically no headway. "Can't I do what I want with my own?" Commodore Vanderbilt had exclaimed, asserting his exclusive right to control the operations of the New York Central system; and that question fairly well represented the popular attitude. That the railroad exercised certain rights of sovereignty, such as that of eminent domain, that it actually used in its operations property belonging to the State, and that these facts in themselves gave the State the right to supervise its management, and even, if necessity arose, to control it — all this may have been recognized as an abstruse legal proposition, but it occupied no practical place in the business consciousness of that time. Naturally the first step of the railroads was therefore to contest the constitutionality of the laws, and while these suits were pending they resorted to various expedients to evade these laws or to mitigate their severity. A touch of liveliness and humor was added to the situation by the thousands of legal fare cases that filled the courts, for farmers used to indulge in one of their favorite agricultural sports — getting on trains and tendering the legal two and a half cents a mile fare, a situation that usually led to ejectment for nonpayment and then to a suit for damages. The railroads easily met the laws forbidding lighter charges for long than for short hauls by increasing the rates for the longer distances, and the laws fixing maximum rates within the State by increasing the rates outside the State. When the courts decided the cases against the railroads, as in most cases they did, these corporations set about to secure the repeal of the laws. They started campaigns of education, frequently through magazine or newspaper articles pointing out the injustice of the Granger laws and insisting that they were working great public damage. It is a fact that a decrease in railroad construction followed the Granger demonstration, and the friends of the railroads insisted that timid capital hesitated to embark in an enterprise that was constantly subject cases sustain practically all the legal contentions made by the Granger legislatures.[1] The cases fixed for all time the point that a State, acting under the police power, may regulate the charges of a railroad even to the extent of fixing maximum rates. They even went so far as to hold that the right to fix rates is not subject to any restraint by the court on the ground of unreasonableness, a principle which the Supreme Court has reversed in more recent times. The courts also held that a State, at least until Congress acted, could regulate interstate commerce, but this decision also has since then been reversed. These subsequent reversals of decisions which were exceedingly popular at the time, however, not only constituted sound law but promoted the public interest, for they established that body of law which has made possible the present more comprehensive system of Federal regulation of railroads.

Meanwhile the demand for regulation was gaining strength in the Eastern States, but for somewhat different reasons. The farmers of New England, New York, and the Eastern region in general had

[1] The cases of particular interest were: Munn vs. Illinois, 94 U. S. 114; Peik vs. Chicago and Northwestern Railway Company, 94 U. S. 164; and Chicago, Burlington and Quincy Railway Company vs. Cutts, 94 U. S. 155.

not particularly sympathized with the Granger legislation; they already had great difficulty in competing with the large Western farms, and a reduction in rates to the seaboard would have made their position even less endurable. This attitude was unquestionably selfish but entirely comprehensible. The agitation for railroad reform in the East came chiefly from the manufacturing and commercial classes. Here the main burden of the complaint was the railroad rebate. This was a method of giving lower rates to large shippers than to small — charging the favored shipper the published rate and then, at stated periods, surreptitiously returning part of the payment. This was perhaps the most vicious abuse of which the railroads have ever been guilty. That the common law forbade the practice and that it likewise violated the implied contract upon which the railroad obtained its franchise was hardly open to dispute; yet up to 1887 no specific law in this country prohibited the practice. For many years the rebate hung over the American business world, a thing whose existence was half admitted, half denied, a kind of ghostly economic terror that seemed persistently to drive the small corporation to bankruptcy and the large corporation to dominating influence. The Standard Oil Company was the "monster" that was believed especially to thrive upon this kind of sustenance, though this was by no means the only industry that maintained such secret relations with the railroads; the Carnegie Steel Corporation, for example, accepted rebates almost as persistently. It was not until 1879, when the Hepburn Committee in New York State had its hearings, that all the facts concerning the rebate were exposed officially to public view. The contracts of the Standard Oil Company with the railroads were placed upon the records and these showed that all the worst suspicions regarding this practice were justified. This disclosure made the railroad rebate one of the most familiar facts in American industrial life; and in consequence a demand arose for Federal legislation that would definitely make the practice a crime and also for some kind of Federal supervision to do effectively the work which the state commissions had failed to do.

By this time it was clear enough that the only hope of adequate regulation lay with the Federal Government. Congressman Reagan, of Texas, had for years been pushing a bill to regulate interstate commerce and to prohibit unjust discriminations by common carriers; other measures had no prerogative to fix rates. Inadequate as this measure seemed to the radical element, it was generally hailed as marking the beginning of an era in the Federal control not only of railroads but of other corporations, and this impression was increased by the high character of the men whom President Cleveland appointed to the first board.

The Interstate Commerce Commission lasted essentially in this form for nearly twenty years. On the whole it was a failure. Such was the judgment passed by Justice Harlan of the United States Supreme Court when he remarked in one of his decisions that the commission was "a useless body for all practical purposes"; and such, indeed, was the judgment of the commission itself, for in its report of 1898 it declared that the attempt at Federal regulation had failed. The chief reasons for this failure, the commission said, were the continued existence of secret rates and the fact that published tariffs were not observed.[1] The managers of the great American railroad systems would not yet admit that the fixing of railroad rates was the concern of any one but themselves, and they

[1] But it should be added that the effectiveness of the commission as an administrative and regulating body was diminished by decisions of the courts, notably the decision of the Supreme Court in the maximum rate case. See 160 U. S. 479.

still regarded railroad management as essentially a private business. If they could obtain large shipments by granting special rates, even though they had to do it by such underhanded ways as granting rebates, they believed that they were entirely justified in doing so. Thus rebates flourished almost as much as ever, passes were still liberally bestowed, and pools were still formed, though they sometimes took the shape of "gentlemen's agreements."

In 1906, when President Roosevelt became intensely active in the railroad problem, conditions were fairly demoralized. Attempts to enforce the anti-pooling clause had led railroads to purchase competing lines, and when the United States Supreme Court pronounced this illegal, the situation became chaotic. The evils of over-capitalization also became an issue of the times. The Interstate Commerce Commission had become almost moribund, and there was a general sentiment that the trouble arose from the fact that the commission had no power to fix rates and that the solution of the railroad problem would come only when such power was vested in it.[1] The Interstate Commerce

[1] The Elkins Act of 1903 had, it is true, increased the effectiveness of the commission in dealing with discriminations, but it had not solved the problem of securing reasonable rates.

Act which became a law on June 29, 1906, was the outcome of one of the greatest battles of President Roosevelt's political life. The act increased the membership of the commission from five to seven members, placed under its jurisdiction not only railroads but pipe lines, express companies, and sleeping-car companies, added to the other familiar restrictions a "commodities clause," which prohibited any railroad from transporting a product which it had produced or mined, "except such articles or commodities as may be necessary and intended for its use in the conduct of its business as a common carrier" — this clause was intended to end the railroad monopoly of the coal mines — and made the failure to observe published tariffs a crime punishable with imprisonment. The amended law did not give the commission the right to fix rates in the first instance but did empower it, on complaint, to investigate charges and on the basis of this investigation to determine just maximum rates, regulations, and practices, though carriers were given the right of appeal to the courts.

Thus, in essence, the public had obtained the reform which it had been demanding for years. The reorganized commission did not hesitate to exercise its new powers. It soon began actually demands for a so-called eight-hour day, and threatened a general strike that would paralyze all business and industry and throw the whole life of the nation into chaos. Properly to appreciate the consequences of this event, it is necessary to keep in mind the fact that the plea for an "eight-hour day" was spurious. An eight-hour day cannot be rigidly enforced on railroads; the workmen well knew this, and indeed they did not really demand such working hours. What they asked for was a full day's pay for eight hours and "time and a half" pay for all in excess of that amount; that is, they demanded an increase in wages. President Wilson, having failed in his attempt to settle the difficulty by arbitration, compelled a Democratic Congress over which his sway was absolute to pass a law — sponsored by Chairman Adamson of the House Committee on Interstate Commerce — which granted practically what the unions demanded. In passing this law, Congress asserted an entirely new power which no one had ever suspected that it possessed — that of fixing the wages which should be paid by common carriers and possibly by other corporations engaged in interstate commerce. The railroads immediately took the case to the United States Supreme Court, which promptly sustained the law. This decision, unquestionably the most radical in the history of that body, declared virtually that Congress could pass any law regulating railroads which the public interest demanded.

And thus, after fifty years of almost incessant struggle with the public, was the mighty railroad monster humbled. It had lost power to regulate the two items which represent the existence of a business — its income and its outgo. The Interstate Commerce Commission was now fixing railroad rates, and Congress was fixing the amounts of railroad wages. It remained for the Great War to precipitate the only logical outcome of this situation — government control. The steadily increasing responsibilities of war soon told heavily upon all lines until, in the latter part of 1917, the whole railroad system of the United States had all but broken down. The unions were pressing demands for wage increases that would have added a billion dollars a year to their annual budgets. The fact that so large a part of the output of American locomotive works was being shipped to the Allies made it difficult for the American lines to maintain their own supply. Nearly all coastwise ships and tugs were utilized for war work, a large part of them had been sent to the other side, and this put an additional strain upon the railroads. The movement of troops, the heavy building operations in cantonments and shipbuilding plants, the manufacture and transportation of munitions, all put an unprecedented pressure upon them. Everywhere there was great shortage of cars, equipment, and materials. Possibly the railroads might have risen to the occasion except for the fact that the enormous increase in the cost of labor and supplies made demands upon their treasuries which they could not meet. They repeatedly asked the Interstate Commerce Commission for an increase in rates, but this request was repeatedly refused. The roads were therefore helpless, and their operations became so congested as to create a positive military danger. Under these circumstances there was profound relief when President Wilson took over the roads and placed them under government control, with William Gibbs McAdoo, Secretary of the Treasury, in active charge.

McAdoo immediately took the step which the Administration, while the railroads were under private control, had steadily refused to sanction, and now increased the rates. These increases were so great that they made the public fairly gasp, but, under the impulse of patriotism, there was a good-natured acquiescence. McAdoo also increased wages by hundreds of millions of dollars. His administration on the whole was an able one. He ignored for the moment the prevailing organization and managed the roads as though they constituted a single system. He instituted economies by concentrating ticket offices, establishing uniform freight classifications, making common the use of terminals and repair shops, abolishing circuitous routes, standardizing equipment, increasing the loads of cars and by introducing a multitude of other changes. All these reforms greatly increased the usefulness of the roads, which now became an important element in winning the war. Properly regarded, the American railroads became as important a link in the chain of communications reaching France as the British fleet itself. It is not too much to say that the fate of the world in the critical year 1918 hung upon this tremendous railroad system which the enterprise and genius of Americans had built up in three-quarters of a century. In February, 1918, Great Britain, France, and Italy made official representations to the American Government, declaring that unless food deliveries could be made as they had been promised by Hoover's food administration, Germany would win the war. McAdoo acted immediately upon this information. He gathered all available cars, taking them away from their ordinary routes, and rushed them from all parts of the country to the great grain producing States. All other kinds of shipments were discontinued; officials and employees from the highest to the lowest worked day and night; and presently the huge supplies of the indispensable food started towards the Atlantic coast. So successful was this operation that, on the 12th of March, the supplies so exceeded the shipping capacity of the Allies that 6318 carloads of food stood at the great North Atlantic ports awaiting transportation. This dramatic movement of American food supplies was an important item in winning the war and fairly illustrated the great part which the American railroads played in turning the tide of battle from defeat to victory.

16

The most comprehensive history of any American railroad system is *The Story of Erie*, by H. S. Mott (1900), but even this is partially unreliable and much of it is compiled from unofficial sources. On the financial history of the Erie Railroad, the really valuable authority is Charles Francis Adams in his *Chapters of Erie* (1871). This book furnishes a full and accurate account of the régime of Daniel Drew, Jay Gould, James Fisk, Jr., and the famous "Erie ring," including "Boss" Tweed, and also throws side lights on the character and career of Commodore Vanderbilt. Among other important histories of particular railroad systems may be mentioned *The Union Pacific Railway*, by John P. Davis (1894) and *History of the Northern Pacific Railroad*, by Eugene V. Smalley (1883); but neither of these volumes covers the recent and more interesting periods in the development of these properties. To get a complete and satisfactory view of the later development of the Northern Pacific system, one must turn to modern biographical works, such as the *Life of Jay Cooke*, by E. P. Oberholtzer (1910), the *Memoirs of Henry Villard* (1909), and the *Life of James J. Hill*, by Joseph Gilpin Pyle (1916), which also recounts at length the rise and development of the Great Northern Railway system. But in these volumes, as in many biographies of great men, the authors often betray a bias and misrepresent facts vital to an understanding of the development of both of these railroad systems. A recent volume entitled the *Life Story of J. P. Morgan*, by Carl Hovey, although extremely laudatory and therefore in many ways misleading, contains valuable information about the development of the Vanderbilt lines after 1880 and also about the financial vicissitudes and rehabilitation of the many Morgan properties, such as the Southern Railway, the modern Erie system, the Northern Pacific, the Reading, and the Baltimore and Ohio.

Some of the railroad companies many years ago themselves published histories of their lines, but most of these attempts were of little value, as they were always too laudatory and one-sided and evidently were usually written for political purposes. The best of this class of railroad histories was a book issued by the Pennsylvania Railroad many years ago, giving a record (largely statistical) of the growth and development of its lines. But this book has been long out of print and covers the period prior to 1885 only.

For original material on American railroad history, one must depend almost entirely on financial and railroad periodicals and official and state documents. By far the most valuable sources for all aspects of railroad building and financing during the long period from 1830 to 1870 are the *American Railroad Journal* (1832–1871) and *Hunt's Merchant Magazine* (1831–1870). Both of these periodicals are replete with details of railroad building and growth. And for the period from 1870 to the present time the best authority is the *Commercial and Financial Chronicle*, with its various supplements. The story of modern railroading is so intertwined with finance and banking that to get any broad and complete view of the subject one must consider it largely from the viewpoint of Wall Street. For facts regarding operation and management of modern railroads, the *Railroad Age-Gazette* also is extremely useful. By far the most valuable sources for railroad statistics, railroad legislation, and all related facts, are the annual reports and bulletins of the Interstate Commerce Commission.

INDEX

247

Harriman, E. H., and Union Pacific, 43, 116, 193 *et seq.*; "community of interest" idea, 43, 44–45, 61, 116–17; and Erie reorganization, 91; Hill and, 150–52, 175–76, 202 *et seq.*; and Central of Georgia Railway, 191, 208; compared with Morgan, 198; death (1909), 208

Harrisburg (Penn.), Pennsylvania constructs line to, 48

Harrison, Fairfax, 119

Hartford (Conn.), railroad in 1840, 10

Helena (Mont.), development of, 131

Hepburn Committee, 231

Hill, J. J., and "community of interest" idea, 61, 116; and Baltimore and Ohio, 115; and Great Northern, 147, 173 *et seq.*; Morgan and, 149–50, 153, 202 *et seq.*; and Northern Pacific, 150, 153; and Harriman, 150–52, 175–76, 202 *et seq.*; growth of Hill lines, 165 *et seq.*; personal characteristics, 166–67; "Hill's Folly," 171; St. Paul, Minneapolis and Manitoba Railroad, 171–173

Hocking Valley Railroad, 149

Holland, St. Paul and Pacific financed in, 170

Hopkins, Mark, 126

Hours of labor, eight-hour day, 237

Hudson and Berkshire Railroad, 25

Hudson River, bridge at Albany, 17

Hudson River Railroad, 12, 27–28, 31; and New York Central, 25; Vanderbilt and, 32, 148

Hudson's Bay Company, Hill's relations with, 167–68

Huntington, C. P., 126, 133, 135, 160, 200–01

Illinois, railroad speculation in, 9; development after 1870, 34; Vanderbilt lines in, 34, 52

Illinois Central Railroad, 119; Federal grant to, 13; Harriman and, 45, 191, 196, 200; extent, 207; and Central of Georgia, 208; Union Pacific and, 210

Indiana, railroad speculation in, 9; development after 1870, 34; Vanderbilt lines in, 34, 52; Chesapeake and Ohio in, 44; Erie development in, 92, 93

Indianapolis, Vanderbilt lines to, 40

Inman, J. H., 186

Interstate Commerce Commission, 209, 232–33, 234–235

Inventions, influence on United States, 1

Iowa, railroads in, 18, 36, 121; Pennsylvania Railroad extends to, 57; "Granger Laws," 224–25; commission established, 228

Jersey City, Pennsylvania Railroad develops line to, 56; Erie directors go to, 82

Jewett, H. J., 88, 89, 90

Kansas, railroads in, 36

Kansas and Pacific Railway, 143

Kansas City, development of, 131

Kansas Pacific Railroad, 130, 132, 155

Kelley, O. H., 219

Kennedy, J. S., 170, 171, 174

King, John, 90

New York Central—*Continued*
21; consolidation to form
(1853), 24–26; connection
with New York City, 30;
further consolidation (1869),
33; state investigation, 37;
Morgan and, 38–39, 45;
growth (1885–93), 40; entry
into New England, 41–42;
"community of interest"
idea applied to, 44, 61;
Harriman and, 45, 206;
in 1919, 46; Union Pacific
and, 208

New York Central and Hud-
son River Railroad, 33,
46

New York, Chicago and St.
Louis Railroad, 35

New York City, effect of Erie
Canal on, 5; railroads in
1840, 10; New York and
Harlem Railroad in, 27;
New York Central and, 29;
railroad approaches to, 30;
Vanderbilt lines in, 40;
monetary center, 47; Pennsylvania Railroad enters,
51, 59; Cassatt and, 58;
real-estate speculation by
Erie Railway, 88; Baltimore
and Ohio connection, 104,
106, 110; rates from Chicago
to, 104; Southern Railway
connection, 191

New York, Lake Erie and
Western Railroad, 89

New York, New Haven and
Hartford Railroad, 45, 60;
see also New Haven

New York, Pennsylvania and
Ohio Railroad, 89

New York, Susquehanna and
Western Railroad, 91–92

New York Tribune on Grange
politics, 223

New York, West Shore and
Buffalo Railroad, 35

"Nickel Plate" route, *see*

New York, Chicago and St.
Louis Railroad

Norfolk and Western Railroad,
61, 116, 182

North Carolina Railroad, 179–
180

Northern Central Railroad, 57

Northern Pacific Railroad, 19;
"community of interest"
idea applied to, 110, 117;
Oregon connections, 133;
competition with Union
Pacific, 136; chartered
(1864), 139; land grant
to, 139–40; Jay Cooke and,
141–42; failure (1873), 142,
166; Villard and, 143–44;
receivers appointed, 146;
plan of merger with Great
Northern, 147–48; reorganization by Morgan, 148–50;
prosperity, 152–53; Hill and,
202, 203; Harriman and,
203–06

Northern Securities Company,
151–52, 176, 204, 205

Northwest, development of,
41

North Western Railroad, *see*
Chicago and North Western
Railroad

Ogdensburg and Lake Cham-
plain Railroad, 42

Ohio, railroad speculation in,
9; effect of railroads in, 34;
Vanderbilt lines in, 34, 52;
Chesapeake and Ohio in,
44; Erie development in,
92, 93

Ohio and Mississippi Railroad,
100

Olcott, F. P., 187

Oregon and Transcontinental
Company, 142, 143, 144,
145

Oregon Railway and Navi-
gation Company, 133, 134,
143